PAINLESS Life Science

Joyce Thornton Barry

Children make our lives worth living!

This book is dedicated to the children in my life:
My beautiful daughter Alanna and handsome son Brendan
My nephews Thomas, Sean, Shane and Zachary and
My nieces, Fiona and Allison

All inquiries should be addressed to:
Barron's Educational Series, Inc.
250 Wireless Boulevard
Hauppauge, NY 11788
www.barronseduc.com

ISBN-13: 978-0-7641-4172-0
ISBN-10: 0-7641-4172-4

Library of Catalog Card No.: 2008042380

Library of Congress Cataloging-in-Publication Data

Barry, Joyce Thornton.
Painless life science / Joyce Thornton Barry.
p. cm.
Includes index.
ISBN-13: 978-0-7641-4172-0
ISBN-10: 0-7641-4172-4
1. Biology—Study and teaching (Secondary) I. Title

QH315.B36 2009
570—dc22

2008042380

Printed in Canada
9 8 7 6 5 4 3

CONTENTS

INTRODUCTION

Life is all around us.

Walking through a field full of flowers, strolling through the forest, playing at the beach, or riding a bicycle after school—all of these events are opportunities for you to observe and appreciate the beauty of life. What makes this earth and us so wonderful is the complexity of life.

You may think that you don't know much about life science, but you will be very surprised at how familiar you already are with the material covered in this book. Life science is also known as biology; it is the study of living things and their environment. Life science is all around you. Try to think of one thing that does not involve life science. It is impossible; all aspects of our world have some connection to life.

HINT

Throughout this book you will see hints. These are little tidbits of information that will help you study, understand, and remember the material being presented. Many words in science come from Latin. If we take the word and translate it from the Latin word to English, we have a better chance of understanding the concept.

The best part of life science is that most concepts can be best described through drawings. Take time and look over the drawings, charts, and diagrams. They too will help you develop a better understanding of the information.

CHAPTER ONE

Life

LIVING VS. NONLIVING

The world around us is made up of many different things, all of which affect others. Everything on the earth can be sorted into two categories; they are considered living or nonliving. **Living things (biotic)** are both similar to and different from each other and from nonliving things. We say that living things are alive. They are often referred to as **organisms**. These organisms are born, grow, change, have the ability to reproduce, and, in the end, die. The changes that an organism goes through over time are called its **life cycle**.

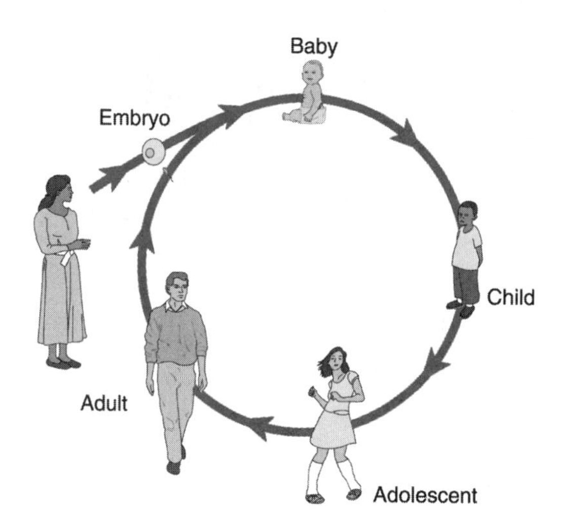

Nonliving things (abiotic) are all around us as well. They are present in nature or can be made by humans. They do not have a life cycle; if they change, they do not change on their own. Nonliving things do not breathe, grow, or die; they are *not alive*. The only way a nonliving thing would be able to change is if something else causes the change. An example of this would be a rock; rocks are nonliving things. A backpack also is nonliving; it was made.

A rock may change over time only if it is exposed to rushing water for a long period of time or if it experiences extreme temperature changes such as ice and heat. These conditions can cause the rock to crack, break, or wear away. The changes that occur to these rocks, or land, is known as **weathering and erosion**. The Grand Canyon is a result of weathering and erosion.

HINT

To determine if something is living, think about how it came to be and how it exists. Was it born? Does it grow? Does it breathe? If it was born, breathes, or grows it is a living thing. If not, it is a nonliving thing.

BRAIN TICKLERS

Set # 1

Place and L or N next to each object to indicate if it is living or nonliving.

1. Tree L
2. Cloud N
3. Bird L
4. Bee L
5. Rock N
6. School N
7. Playground swingset N
8. Flower L

9. List the stages of the human life cycle based on this diagram.

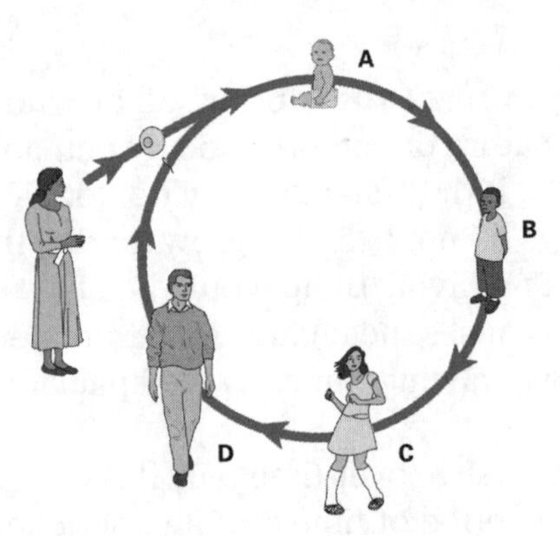

Begin with Mother and Father create A → B → C → D

10. Explain how this rock may have gotten holes and crevices. It is a change that happened over a long period of time.

(Answers are on page 27.)

BASIC NEEDS OF LIVING THINGS

Animals are living; they need air, water, and food to live and thrive.

- Animals take in **air** by breathing. They need **oxygen** from the air; it allows the animal to make and use *energy*, which it needs to survive.

HINT

Think of the shape of your mouth when it is open (the O shape). This should remind you that we need O (oxygen) to survive.

- **Water** is also necessary for the animal to survive. Animals use water to break down and move materials throughout their bodies.
- Animals cannot make their own **food** so they must eat to get **nutrients**. The nutrients that they get from food are necessary for them to grow and have energy.

Plants are also living; they need air, water, nutrients, and light to live.

- Plants take in **air** through their leaves. The part of the air that plants take in and use is **carbon dioxide**. This is necessary for the plant to create food. Plants can make their own food by a process called photosynthesis.
- **Water** is used by the plant to move materials up the plant and to create food.
- **Nutrients** from the soil enter the plant through the roots. They are necessary for the plant to survive.
- **Light** is the most important thing for all plants. Light gives the plant the *energy* it needs to survive.

HINT

Whenever the sun is mentioned, think of energy. Most energy on the earth starts with the energy from the sun.

Interaction between living and nonliving things

Everything on earth exists together, and the result is a beautiful balance. Living things depend on nonliving things and each other to survive, which is why we often say we live off the earth.

The basic needs for all living things are energy, water, gases such as oxygen or carbon dioxide, and minerals. Plants and animals help each other to survive. The plant needs carbon dioxide and gives off oxygen; animals need oxygen and give off carbon dioxide.

HINT

Think of a circle with a human exhaling carbon dioxide and a plant taking it in and the plant giving off oxygen with the human taking it in.

Fossils

Sometimes organisms that lived at one time leave an imprint in the ground after they die. When these organisms decay, they leave an indentation similar to what your shoe would leave on muddy ground.

The remnant of an organism that once lived is called a **fossil**. If an organism gets trapped in soil, it will die, and over time a rock may form around it. Many years later, the fossil may be discovered when the rock is exposed or broken apart and the mold of the once living organism is exposed. Thanks to fossils, scientists are able to tell us about animals that lived long ago. This is how they have been able to collect so much information about our past and especially the history of dinosaurs. Their skeletons were found, helping scientists create the story of the dinosaurs' lives on earth. They have also shown us how organisms change over time.

HINT

How do we figure out if an object is living or nonliving? Ask the question, Does it take in and give off any gases? If it does, then it is living.

BRAIN TICKLERS
Set # 2

1. List three things that you need to live and thrive:
   water ______ food ______ oxygen ______

2. Water helps living things suvive by d ______.
   - a. bringing in gases
   - b. keeping them clean
   - c. creating food
   - d. moving materials

3. Air and water are necessary for __________ to survive.
   - a. plants
   - b. animals
   - (c) both plant and animals
   - d. nonliving things

4. A fossil ______ is an imprint of a once-living organism.

5. A mountain is a __________.
   - (a) nonliving thing that occurs in nature
   - b. nonliving thing that is man-made
   - c. living thing that occurs in nature
   - d. living thing that is man-made

6. List two nonliving things that all living things need to survive.

7. Everything on the earth exists together to result in a beautiful balance. Explain what is meant by this statement.

(Answers are on page 28.)

LIFE FUNCTIONS

Living things are very complicated organisms. Throughout their lives, they perform specific life functions. These life functions help organisms survive in their changing environment.

The **life functions** are

- Regulation
- Transport
- Reproduction
- Growth and development
- Nutrition
- Respiration
- Locomotion

Each life function has a specific role in helping organisms live and adjust to their ever-changing surroundings. The life functions help organisms maintain **homeostasis**. Homeostasis is how organisms keep their bodies stable. They do this by responding to their external and internal environments. Here are two examples of homeostasis:

- You sweat in warm weather to cool off your body and shiver in cold weather to warm up your body.
- Your heartbeat gets faster when you run a race. Your heart does this so that more blood will flow through your body bringing more oxygen to your cells.

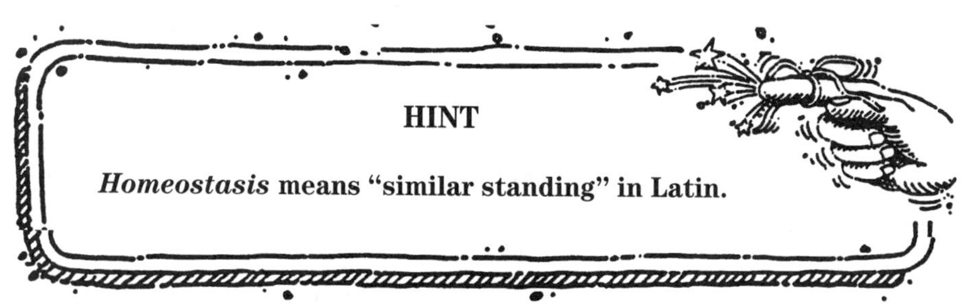

HINT

***Homeostasis* means "similar standing" in Latin.**

The basic unit of all living things is the cell. All organisms are made up of cells. Some organisms are made up of only one cell; we call these organisms **unicellular**. Examples of unicellular organisms are bacteria and protozoa. The majority of living things are made up of many cells, like people; they are called **multicellular**. Animals are multicellular. All living things go through the life functions no matter how many cells they are made of.

Levels of organization

In unicellular (single-celled) organisms, the single cell performs all life functions. It functions independently. However, multicellular (many-celled) organisms have various levels of organization within them. Individual cells may perform specific functions and also work together for the good of the entire organism. The level of organization ranges from simplest to most complex: Each item represents a different level of organization. Organelles, for example, operate only within a cell, and what they do has a direct effect only on the cell. Many cells that work together to perform a specific function in the body, work together to form tissue. A tissue, such as muscle, operates on a higher level of organization. Two or more tissue types occur in organs, which represent yet a higher level of organization.

The levels of organization can be summarized as

atom → molecule → organelle → cell → tissue → organ → organ system → organism

BRAIN TICKLERS
Set # 3

1. An organism responds to its external and internal environments by maintaining __________.
2. Choose the correct level of organization:
   a. organism → cells → tissues → organs → organ systems
   b. cells → tissues → organs → organ systems → organism
   c. organism → organ system → organs → tissues → cells
   d. tissues → organ systems → organs → organisms → cells

(Answers are on page 28.)

Let's explore each of the life functions.

Regulation

All organisms have developed systems in their body to help them *respond to change* in their environment. Anything an organism responds to is known as a **stimulus**. A stimulus could be a smell, sound, chemical, temperature change, or even another organism. Organisms are able to respond to stimuli so that they can keep their bodies stable—homeostasis. What the organism does because of the stimuli is called a **response**. Organisms make changes over time to help them live in a certain environment; these changes are called **adaptation**.

- Animals that live in cold regions have thick fur and a layer of fat. This is an adaptation that helps them keep warm in the cold weather.
- The beaks of birds are shaped to help them eat specific foods from their environment. Pelicans have a big sack in their bill to hold fish. Hummingbirds have long skinny beaks so they can get into the tiny spaces of flowers.

Transport

Transport relates to how the organism is able to move materials throughout its system. Materials such as nutrients, water, and air enter the organism and move through the transport system to the cells where they are used. The transport system then takes the waste out of the cells and delivers it to the organ where it will be disposed of. For example, the circulatory system is the transport system in animals. Nutrients pass from food into the digestive tissues, but then they need to get to the cells of the body. The blood picks up the nutrients to transport them to cells of the body. The blood also takes the waste from the cells and delivers it to the appropriate organs for disposal.

Reproduction

All living things are able to produce **offspring**; we usually call them babies. Some offspring are born looking like their parent. Some are born as one form and change over time into an adult form, such as a butterfly.

There are two types of reproduction that happen in living things. **Asexual reproduction** occurs in unicellular organisms. The new offspring is created from *one parent*. This offspring looks identical to the parent. **Sexual reproduction** is when *two parents'* sex cells join to form a new organism that has a combination of both parents and is a new individual.

Growth and development

Most animals are born as a smaller version of what an adult looks like. As they grow, their bones get longer and bigger; they grow more fur or hair; and they are able to take care of themselves by getting food, shelter, and water. Cell reproduction and development help all organisms grow.

Some animals are born as one type of organism, and as they grow, they then go into a changing stage called **metamorphosis**. Tadpoles change into frogs, caterpillars change into butterflies, and mealworms turn into beetles. This is their way of growing and changing.

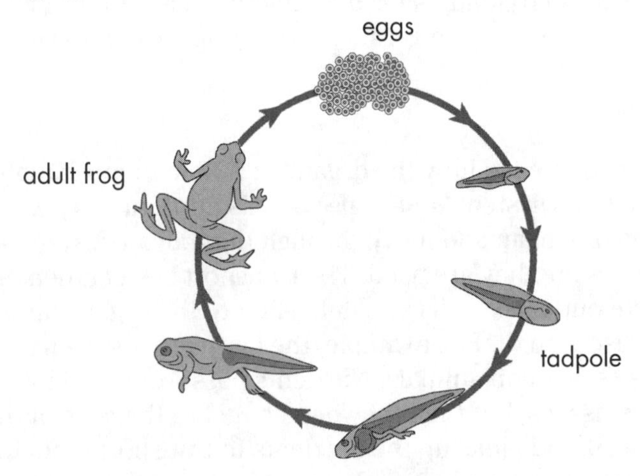

Frog life cycle

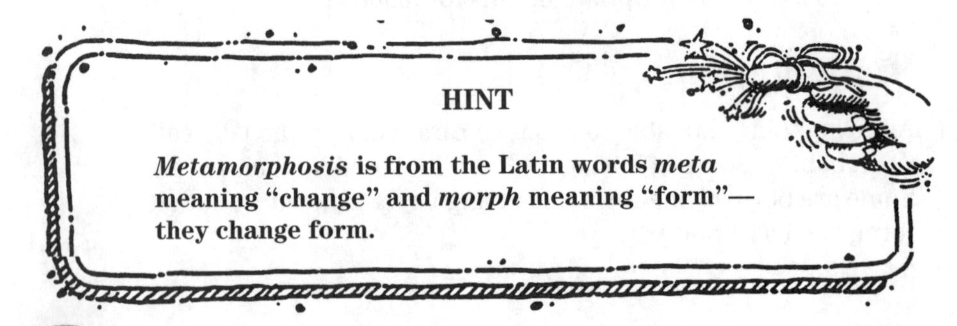

HINT

***Metamorphosis* is from the Latin words *meta* meaning "change" and *morph* meaning "form"—they change form.**

Plants can grow from a seed into a flower, tree, or bush. Plants reproduce by producing flowers and fruits that have seeds; the seeds then grow into plants. Other plants develop outgrowths that can grow into adult plants.

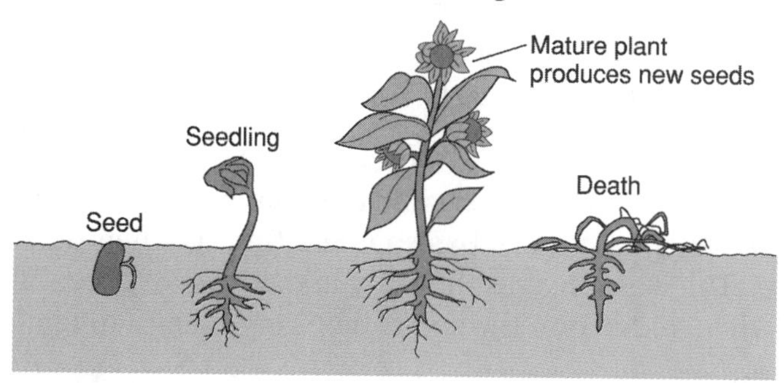

Nutrition

All living things need nutrients to survive. Animals take in food as a nutrient to give them energy and help them grow. Plants get their nutrients from the soil; this helps them grow and perform photosynthesis. **Photosynthesis** is the process by which plants can make their own food.

Organisms get nutrients in one of two ways:

Autotrophs—able to make their own food
Heterotrophs—cannot make their own food

HINT

Think of the Latin to remember *autotrophs* and *heterotrophs*.

- ***troph*—food, nourishment**
- ***auto*—self**
- ***hetero*—other**
- ***hetero troph*—others for food**
- ***auto troph*—self for food**

Nutrition can be broken down into three phases:

Ingestion—The food and nutrients are taken in. This is commonly called eating.

Digestion—The food moves into the digestive system and is broken down into useful chemicals that the organism can use to continue its life functions.

Absorption—The nutrients are absorbed by the villi of the small intestine for use in the body.

Organisms get rid of the unused food by **egestion (elimination)**. This is also called eliminating waste: All animals get rid of solid waste after passing food through their digestive systems. The waste leaves the body in the form of **feces**. This waste when mixed into the soil can be very helpful to plants growing.

Animals get rid of their gaseous waste by exhaling carbon dioxide through their mouths and noses; this is part of the life function called respiration.

Plants get rid of their metabolic (chemical) waste through their roots. Plants get rid of gaseous waste through their leaves in the form of oxygen.

HINT

Here is an easy way to remember how food is moved through the body.

- **Ingest—take in**
- **Digest—break down**
- **Absorb—soak up**
- **Egest or eliminate—get rid of**

Respiration (breathing)

All living things do some type of breathing. Most animals take in oxygen through their mouths; fish breathe through their gills. They **inhale** (take in) oxygen, which they use to create energy. Oxygen is found in the air all around us. Animals **exhale** (give off) carbon dioxide, which is used by plants. The plants take in carbon dioxide and give off oxygen.

HINT

Here is an easy way to remember the steps in respiration.

- **Inhale—take in air/gases**
- **Exhale—let out air/gases (This is an exit.)**

Locomotion

Locomotion is the life function that all living things need to perform to survive. Locomotion occurs when an organism moves to obtain food, water, shelter, sunlight, or to protect itself. Some unicellular organisms move by little hairs on their outside called cilia or one long hair called a flagellum. They move around to obtain food and to avoid predators. Plants don't move from one place to another, but their leaves do move toward the sun to get light so that they can perform photosynthesis. Animals have bones, muscles, ligaments, and tendons that help them to move from one place to another.

BRAIN TICKLERS
Set # 4

1. The life process where animal bones get longer and bigger is called __________.
   a. growth
   b. reproduction
   c. breathing
   d. nutrition

2. Frogs and butterflies look very different from when they are born. The changes they go through are called __________.
   a. skipping
   b. metamorphosis
   c. fading
   d. breathing
3. What materials make up fecal waste (feces)?
4. __________ are able to make their own food by a process called photosynthesis.
5. A seed growing into a plant is a form of which life process?
6. Match the life function with the need.

a.	Locomotion	1.	Breathing
b.	Transport	2.	Respond to stimuli
c.	Regulation	3.	Materials move in organism
d.	Growth and development	4.	Ingest, digest, egest
e.	Respiration	5.	Movement to get shelter
f.	Reproduction	6.	Metamorphosis
g.	Nutrition	7.	New offspring

7. Choose a life function and explain why it is necessary for a living thing to survive.
8. An autotroph is able to __________.

(Answers are on page 28.)

BIOCHEMISTRY

Life science begins with the creative combination of materials to form new organisms. In order to understand and appreciate how amazing life science is, we have to explore the world of biochemistry. **Biochemistry** is the study of chemical substances and the process that occurs in living organisms.

Before we can learn about the chemistry, we have to understand what all things are made of and how they are put together. When we discuss chemicals, air, organisms, rocks, or molecules we can call all of them **matter**. Matter is anything that has mass and takes up space. All things that can be measured are made up of matter. All matter is made up of **atoms**, which are far too small to see. Matter in its smallest form are called atoms. Atoms are often called the *building blocks* and are the basis for everything in the universe. Atoms are constantly moving; the more energy they have, the faster they move. Because atoms are so small that we can't see them, scientists have drawn models of what the atom is made of. If we were able to take an atom and look at its parts it would look like a bull's-eye target.

In the very center of the atom is a **nucleus**. The nucleus contains two kinds of particles called protons and neutrons. A **proton** has a positive charge, and the **neutron** has no charge. Surrounding the nucleus of the atom is the **electron cloud**. The **electrons** have a negative charge. The electrons are always found whizzing around the center in areas called **orbitals**. Atoms can be drawn different ways, but they always have the nucleus in the middle and the orbital, also known as the electron clouds, on the outside.

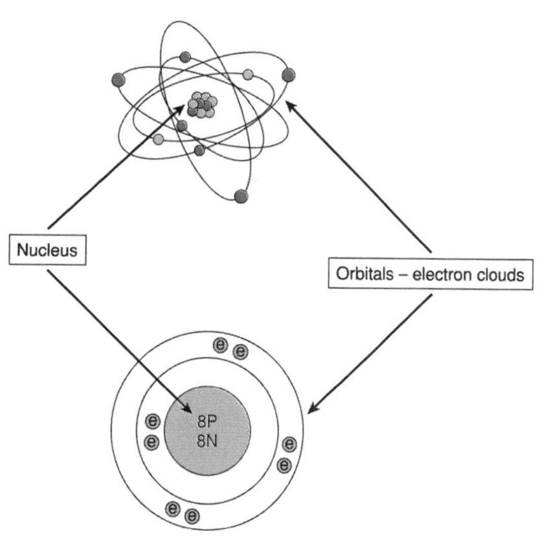

All atoms have energy. Since the atom is constantly moving, it has energy; the amount of energy results from how fast the atom's electrons are moving in the orbital.

Atomic Structure

Name of Part	Location	Charge	Characteristics
Proton	Nucleus	Positive	Has mass
Electron	Outside of nucleus in orbital/electron cloud	Negative	Energy– no mass
Neutrons	Nucleus	No charge	Has mass

HINT

All atoms have the same parts. An easy way to remember the parts is to think of the word

PEN—Protons, Electrons, Neutrons.
Protons—positive charge
Electrons—electricity, negative charge
Neutrons—neutral, no charge

Since all matter is made up of atoms, what makes them different? What makes matter different from each other is the type of atoms that they are made up of. Anything that is made up of only one type of atom is called an **element**. There are over 100 elements; they are different from each other based on the number of protons they have in their nuclei.

Here are some common elements that you may be familiar with. The hydrogen atom is the simplest of all atoms.

Hydrogen (H) has one proton and one electron.

Helium (He) has two protons, two neutrons, and two electrons.

Oxygen (O) has eight protons, eight neutrons, and eight electrons.

Carbon (C) has six protons, six neutrons, and six electrons.

BRAIN TICKLERS
Set # 5

1. The center of the atom is called the ___________.
2. Electrons are found in an atom's ___________.

a. nucleus
b. orbital
c. center
d. air

3. Label this atom.

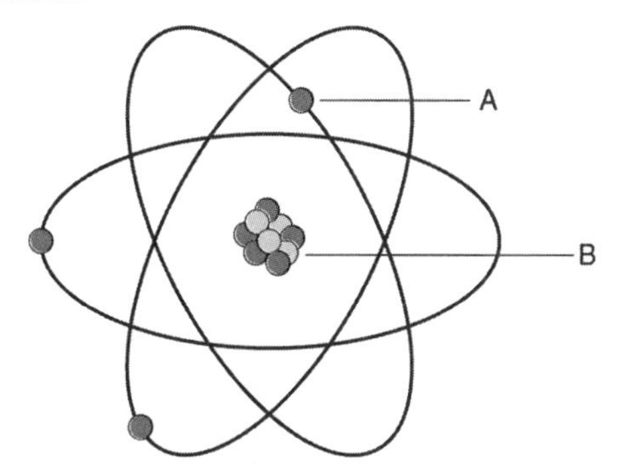

4. Describe how elements differ from each other.
5. Matter is anything that has __________ and takes up __________.

(Answers are on page 29.)

Periodic table of elements

It would be very difficult to remember the properties of each element. To make life easier, chemists have organized the elements based on their different properties; how many protons they have, the type of substance they make, and how they react to other chemicals. The elements are arranged on a chart called the **periodic table of elements**.

Each element has a name; it is noted on the periodic table of elements by one or two letters. Here are some examples:

H, hydrogen	K, potassium
C, carbon	O, oxygen
He, helium	Al, aluminum
Fe, iron	N, nitrogen

In the periodic table, the elements are actually arranged in order of increasing **atomic number**—the number of protons in one atom of a particular element. An undisturbed atom is called electrically neutral, which means that the number of electrons in the atom is the same as its atomic number.

PERIODIC TABLE

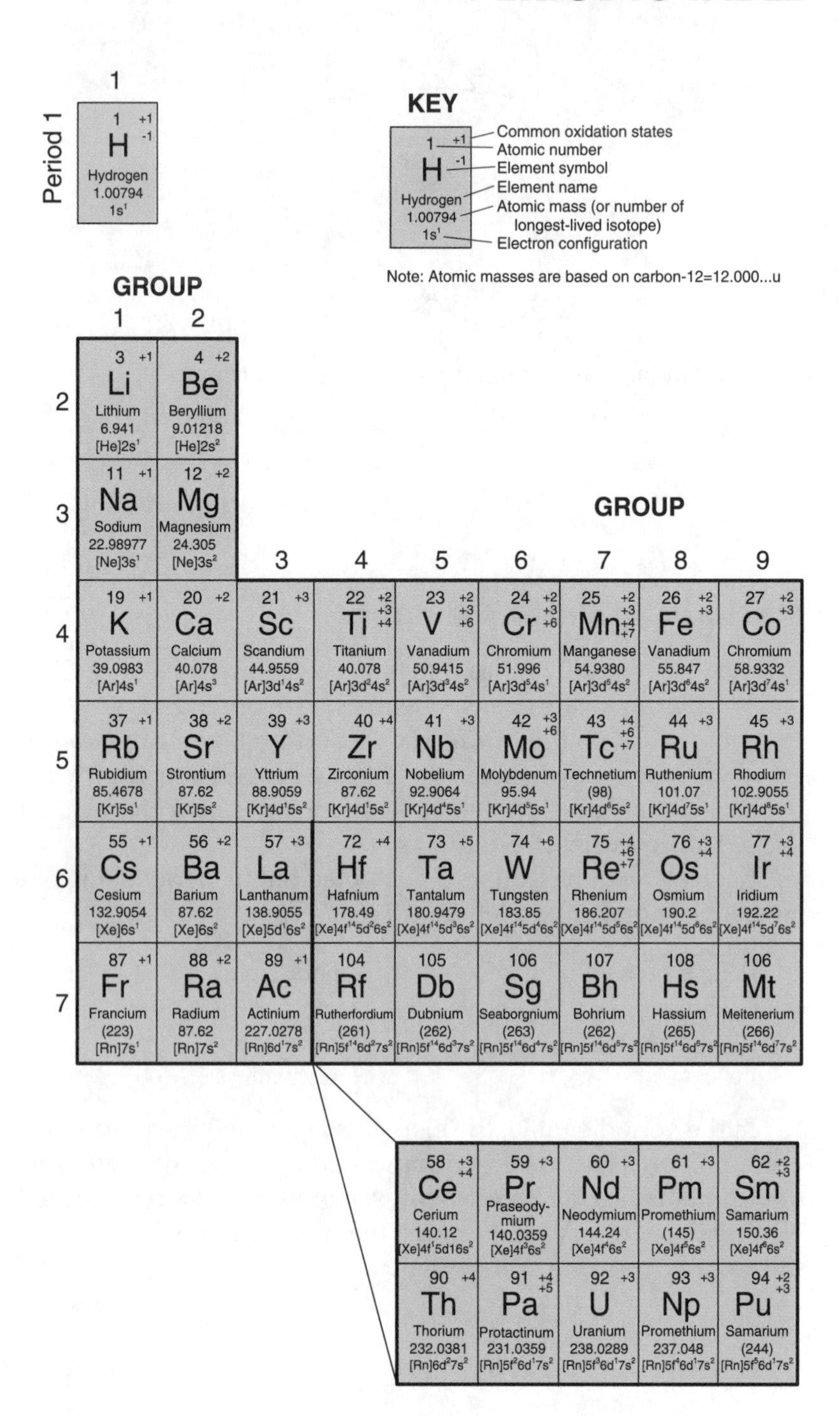

KEY

1 +1 -1 H Hydrogen 1.00794 1s^1	
+1	Common oxidation states
1	Atomic number
H	Element symbol
Hydrogen	Element name
1.00794	Atomic mass (or number of longest-lived isotope)
1s^1	Electron configuration

Note: Atomic masses are based on carbon-12=12.000...u

Period	GROUP 1	2	3	4	5	6	7	8	9
1	1 +1 -1 H Hydrogen 1.00794 1s^1								
2	3 +1 Li Lithium 6.941 [He]2s^1	4 +2 Be Beryllium 9.01218 [He]2s^2							
3	11 +1 Na Sodium 22.98977 [Ne]3s^1	12 +2 Mg Magnesium 24.305 [Ne]3s^2							
4	19 +1 K Potassium 39.0983 [Ar]4s^1	20 +2 Ca Calcium 40.078 [Ar]4s^3	21 +3 Sc Scandium 44.9559 [Ar]3d^{1}4s^2	22 +2 +3 +4 Ti Titanium 40.078 [Ar]3d^{2}4s^2	23 +2 +3 +6 V Vanadium 50.9415 [Ar]3d^{3}4s^2	24 +2 +3 +6 Cr Chromium 51.996 [Ar]3d^{5}4s^1	25 +2 +3 +4 +7 Mn Manganese 54.9380 [Ar]3d^{5}4s^2	26 +2 +3 Fe Vanadium 55.847 [Ar]3d^{6}4s^2	27 +2 +3 Co Chromium 58.9332 [Ar]3d^{7}4s^1
5	37 +1 Rb Rubidium 85.4678 [Kr]5s^1	38 +2 Sr Strontium 87.62 [Kr]5s^2	39 +3 Y Yttrium 88.9059 [Kr]4d^{1}5s^2	40 +4 Zr Zirconium 87.62 [Kr]4d^{1}5s^2	41 +3 Nb Nobelium 92.9064 [Kr]4d^{4}5s^1	42 +3 +6 Mo Molybdenum 95.94 [Kr]4d^{5}5s^1	43 +4 +6 +7 Tc Technetium (98) [Kr]4d^{6}5s^2	44 +3 Ru Ruthenium 101.07 [Kr]4d^{7}5s^1	45 +3 Rh Rhodium 102.9055 [Kr]4d^{8}5s^1
6	55 +1 Cs Cesium 132.9054 [Xe]6s^1	56 +2 Ba Barium 87.62 [Xe]6s^2	57 +3 La Lanthanum 138.9055 [Xe]5d^{1}6s^2	72 +4 Hf Hafnium 178.49 [Xe]4f^{14}5d^{2}6s^2	73 +5 Ta Tantalum 180.9479 [Xe]4f^{14}5d^{3}6s^2	74 +6 W Tungsten 183.85 [Xe]4f^{14}5d^{4}6s^2	75 +4 +6 +7 Re Rhenium 186.207 [Xe]4f^{14}5d^{5}6s^2	76 +3 +4 Os Osmium 190.2 [Xe]4f^{14}5d^{6}6s^2	77 +3 +4 Ir Iridium 192.22 [Xe]4f^{14}5d^{7}6s^2
7	87 +1 Fr Francium (223) [Rn]7s^1	88 +2 Ra Radium 87.62 [Rn]7s^2	89 +1 Ac Actinium 227.0278 [Rn]6d^{1}7s^2	104 Rf Rutherfordium (261) [Rn]5f^{14}6d^{2}7s^2	105 Db Dubnium (262) [Rn]5f^{14}6d^{3}7s^2	106 Sg Seaborgnium (263) [Rn]5f^{14}6d^{4}7s^2	107 Bh Bohrium (262) [Rn]5f^{14}6d^{5}7s^2	108 Hs Hassium (265) [Rn]5f^{14}6d^{6}7s^2	106 Mt Meitenerium (266) [Rn]5f^{14}6d^{7}7s^2

58 +3 +4 Ce Cerium 140.12 [Xe]4f^{1}5d16s^2	59 +3 Pr Praseody-mium 140.0359 [Xe]4f^{3}6s^2	60 +3 Nd Neodymium 144.24 [Xe]4f^{4}6s^2	61 +3 Pm Promethium (145) [Xe]4f^{5}6s^2	62 +2 +3 Sm Samarium 150.36 [Xe]4f^{6}6s^2
90 +4 Th Thorium 232.0381 [Rn]6d^{2}7s^2	91 +4 +5 Pa Protactinum 231.0359 [Rn]5f^{2}6d^{1}7s^2	92 +3 U Uranium 238.0289 [Rn]5f^{3}6d^{1}7s^2	93 +3 Np Promethium 237.048 [Rn]5f^{4}6d^{1}7s^2	94 +2 +3 Pu Samarium (244) [Rn]5f^{5}6d^{1}7s^2

OF THE ELEMENTS

18
2 0 **He** Helium 4.00260 $1s^2$

GROUP

10	11	12	13	14	15	16	17	18
			5 +3 **B** Boron 10.81 $[He]2s^22s^1$	6 -4 +2 +4 **C** Carbon 12.011 $[He]2s^22s^2$	5 -3 -2 -1 +1 +2 +3 +4 +5 **N** Nitrogen 14.0067 $[He]2s^22s^3$	6 -2 **O** Oxygen 15.9994 $[He]2s^22s^4$	5 +3 **F** Flourine 18.998403 $[He]2s^22s^5$	6 -4 +2 +4 **Ne** Neon 20.1797 $[He]2s^22s^6$
			13 +3 **Al** Aluminum 26.98154 $[Ne]3s^23p^1$	14 -4 +2 +4 **Si** Silicon 28.0855 $[Ne]3s^23p^2$	15 -3 +3 +5 **P** Phosphorus 30.97376 $[Ne]3s^23p^3$	16 -2 +4 +6 **S** Sulfur 32.066 $[Ne]3s^23p^4$	17 -1 +1 +3 +5 +7 **Cl** Chlorine 35.453 $[Ne]3s^23p^5$	18 0 **Ar** Argon 39.948 $[Ne]3s^23p^6$
28 +2 +3 **Ni** Nickel 58.69 $[Ar]3d^84s^2$	29 +2 +3 **Cu** Copper 63.546 $[Ar]3d^{10}4s^1$	30 +2 **Zn** Zinc 65.93 $[Ar]3d^{10}4s^2$	31 +2 +3 **Ga** Gallium 69.72 $[Ar]3d^{10}4s^24p^1$	32 +2 +3 **Ge** Germanium 72.61 $[Ar]3d^{10}4s^2 4p^2$	33 +2 +4 +5 **As** Arsenic 74.9216 $[Ar]3d^{10}4s^2 4p^3$	34 -2 +4 +6 **Se** Selenium 78.96 $[Kr]3d^{10}4s^25p^4$	35 -1 +1 +5 **Br** Bromine 121.757 $[Kr]3d^{10}4s^25p^5$	36 0 +2 **Kr** Krypton 121.757 $[Kr]3d^{10}4s^25p^6$
46 +2 +4 **Pd** Palladium 106.42 $[Kr]4d^{10}$	47 +1 **Ag** Siver 107.8682 $[Kr]4d^{10}5s^1$	48 +2 **Cd** Cadnium 112.41 $[Kr]4d^{10}5s^2$	49 +3 **In** Indium 114.82 $[Kr]4d^{10}5s^25p^1$	50 -4 +2 +4 **Sn** Tin 118.710 $[Kr]4d^{10}5s^25p^2$	51 -3 +3 +5 **Sb** Antimony 121.757 $[Kr]4d^{10}5s^25p^3$	52 -2 +4 +6 **Te** Tellurium 121.757 $[Kr]4d^{10}5s^25p^4$	53 -1 +1 +5 +7 **I** Iodine 121.757 $[Kr]4d^{10}5s^25p^5$	54 0 +2 +4 +6 **Xe** Xenon 121.757 $[Kr]4d^{10}5s^25p^6$
78 +2 +4 **Pt** Platinum 195.08 $[Xe]4f^{14}5d^96s^1$	79 +1 +3 **Au** Gold 196.9665 $[Xe]4f^{14}5d^{10}6s^1$	80 +1 +2 **Hg** Mercury 200.59 $[Xe]4f^{14}5d^{10}6s^2$	81 +1 +3 **Ti** Thallium 204.383 $[Xe]4f^{14}5d^{10} 6s^26p^1$	82 +2 +4 **Pb** Lead 207.2 $[Xe]4f^{14}5d^{10} 6s^26p^2$	83 +3 +5 **Bi** Bismuth 208.9804 $[Xe]4f^{14}5d^{10} 6s^26p^3$	84 +2 +4 **Po** Potassium (209) $[Xe]4f^{14}5d^{10} 6s^26p^4$	85 **At** Astatine (210) $[Xe]4f^{14}5d^{10} 6s^26p^5$	86 0 **Rn** Radon (222) $[Xe]4f^{14}5d^{10} 6s^26p^6$
110 **Uun** Ununnilium (269) $[Rn]4f^{14}6d^87s^2$	111 **Uuu** Unununium (272) $[Rn]5f^{14}6d^97s^2$	112 **Uub** Ununbium (277) $[Rn]5f^{14}6d^{10}7s^2$						

63 +2 +3 **Eu** Europium 151.96 $[Xe]4f^76s^2$	64 +3 **Gd** Gadolinium 157.25 $[Xe]4f^75d^16s^2$	65 +3 **Tb** Terbium 158.9254 $[Xe]4f^96s^2$	66 +3 **Dy** Dysprosium 162.50 $[Xe]4f^{10}6s^2$	67 +3 **Ho** Holmium 164.9304 $[Xe]4f^{11}6s^2$	68 +3 **Er** Erbium 167.26 $[Xe]4f^{12}6s^2$	69 +3 **Tm** Thulium 168.9342 $[Xe]4f^{13}6s^2$	70 +2 +3 **Yb** Ytterbium 173.04 $[Xe]4f^{14}6s^2$	71 +3 **Lu** Lutetium 174.967 $[Xe]4f^{14}5d^16s^2$
95 +4 **Am** Americium (243) $[Rn]5f^77s^2$	96 +3 **Cm** Curium (247) $[Rn]5f^76d^17s^2$	97 +3 +4 **Bk** Berkelium (247) $[Rn]5f^97s^2$	98 +3 **Cf** Californium (251) $[Rn]5f^17s^2$	99 **Es** Einsteinium (252) $[Rn]5f^{11}7s^2$	100 **Fm** Fermium (257) $[Rn]5f^{12}7s^2$	101 **Md** Mendelevium (258) $[Rn]5f^{13}7s^2$	102 **No** Nobelium (259) $[Rn]5f^{14}7s^2$	103 **Lr** Lawrenncium (262) $[Rn]5f^{14}6d^17s^2$

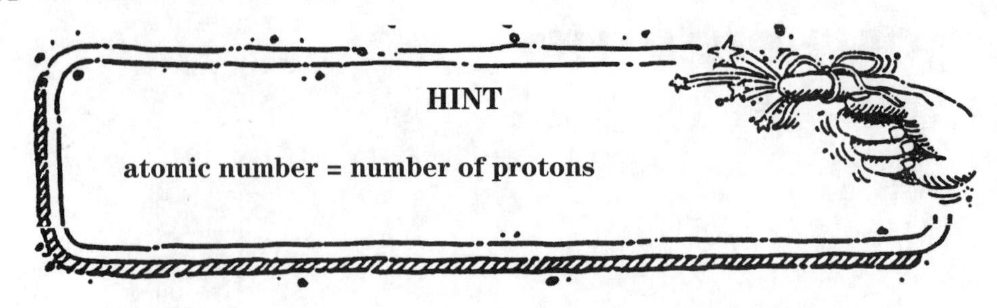

Another important property used to describe elements is **atomic mass**. The atomic mass is the mass of an atom (also known as the atomic weight). Protons and neutrons have mass and thus have weight. The atomic mass is how much the protons and neutrons in the element weigh.

If you look at how the elements are organized on the periodic table, you will notice the atoms get heavier as you go down a column or to the right across a row. This helps scientists understand how certain elements will behave when combined with others.

Elements that are found most in living things are

N, nitrogen
O, oxygen
C, carbon
H, hydrogen

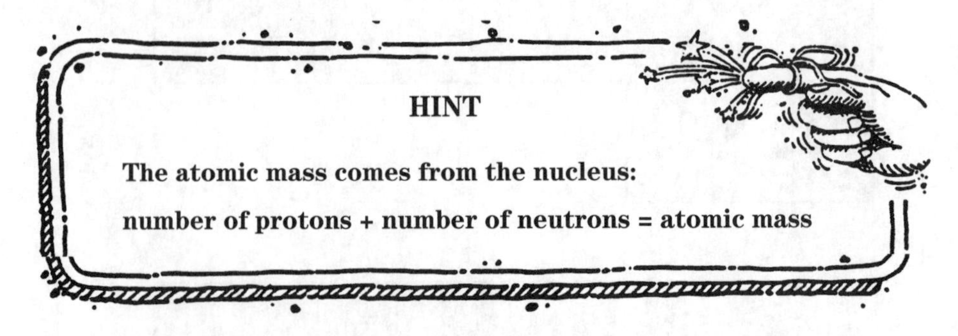

So now that we know the basics of what an atom is, what it looks like, and how elements are different from each other, we are ready to explore molecules and compounds. How are they created? **Molecules** are formed when two atoms are combined. A **compound** is a molecule that contains at least two different elements. All compounds are molecules, but not all molecules are compounds.

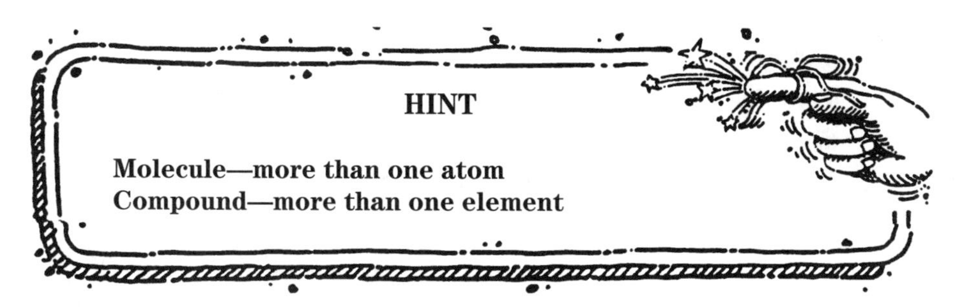

HINT

Molecule—more than one atom
Compound—more than one element

Water (H_2O), sodium chloride or salt (NaCl), carbon dioxide (CO_2), and methane (CH_4) are compounds because each is made from more than one element. The numbers noted next to each letter shows how many atoms of each element are needed to create the item.

Water or H_2O has two hydrogen atoms and one oxygen atom.

Sugar or $C_6H_{12}O_6$ has six carbon atoms, twelve hydrogen atoms, and six oxygen atoms.

BRAIN TICKLERS
Set # 6

1. If you were to take an object—any object—and break it into its smallest part, you would have the __________.
2. Two or more atoms with the same atomic number are called __________.
   a. compounds
   b. neurons
   c. molecules
   d. elements

3. Salt is made up of sodium (Na) and chlorine (Cl) atoms; it is considered to be a __________.
   a. compound
   b. neuron
   c. molecule
   d. element
4. All elements are organized on a large chart called the __________.

(Answers are on page 29.)

Organic vs. inorganic

All compounds in living things are divided into two different categories: **organic compounds** and **inorganic compounds**. Organic compounds contain both carbon and hydrogen. Inorganic compounds can have carbon or hydrogen or neither.

Inorganic—H_2O, CO_2, NaCl
Organic—sugar ($C_6H_{12}O_6$), methane gas (CH_4)

Organic compounds are essential for all living things. They can be divided into four groups: carbohydrates, lipids, proteins, and nucleic acids.

Carbohydrates are the energy compounds. They are made up of carbon, hydrogen, and oxygen. When we eat fruits, bread, pasta, and vegetables, we are storing energy for later use.

Lipids are the protection and storage compounds. Lipids are made up of carbon, hydrogen and oxygen, in the form of fatty acids and glycerol. Animals get lipids from eating oils, nuts, butter, and the fat in meats.

Proteins are the growth and repair compounds. They all contain carbon, hydrogen, oxygen, and nitrogen. They make up muscles and enzymes. Animals get protein from meat, dairy, and fish products.

Nucleic acids are found in all cells. They make up genetic information called deoxyribonucleic acid, commonly known as DNA. It is made up of carbon, hydrogen, oxygen, and phosphorus. The organization of the nucleic acids results in organisms having their own genetic individuality.

Organic Compounds

	Carbohydrate	Lipids	Protein	Nucleic Acids
Composition	Carbon, Hydrogen, and Oxygen — $C_6H_{12}O_6$	Carbon, Hydrogen, and Oxygen	Carbon, Hydrogen, Oxygen, and Nitrogen	Carbon, Hydrogen, Oxygen, Nitrogen, and Phosphorus
Common Name	Sugar, starch, cellulose, glucose, fructose	Fat, oil, wax	Amino acids, enzymes	DNA
Function	Releases energy when broken down, energy for all life functions	Storage of energy, insulation	Builds cells, muscles, etc.	Carries all genetic information for heredity
Location	Fruit, plant walls, cell membranes	Cell membranes	Meat, small particles in cell membrane	DNA in chromosomes, mitochondria, chloroplasts, and nucleus

HINT

Organic—both carbon and hydrogen
Inorganic—either carbon or hydrogen or neither
Carbohydrate—sugar, energy
Lipids—fat, storage, insulation
Protein—growth and repair
Nucleic Acids—DNA-genetic information

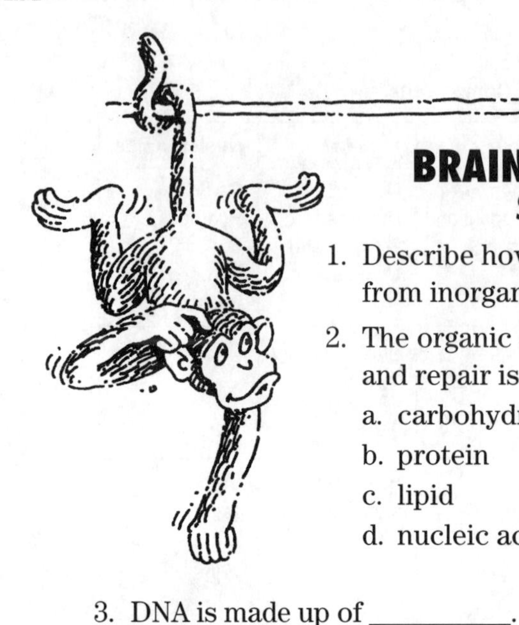

BRAIN TICKLERS
Set # 7

1. Describe how organic compounds differ from inorganic compounds.
2. The organic compound used for growth and repair is the __________.
   a. carbohydrate
   b. protein
   c. lipid
   d. nucleic acid
3. DNA is made up of __________.
   a. carbohydrate
   b. protein
   c. lipid
   d. nucleic acid
4. The organic compound that gives energy is _________.
   a. carbohydrate
   b. protein
   c. lipid
   d. nucleic acid
5. Water is made up of the elements __________ and __________, which make it a(n) __________ compound.
6. Cell membranes are made up of __________.
7. Describe how carbon is such an important element.

(Answers are on page 29.)

WRAPPING UP

- Living things depend on and interact with nonliving things for their survival.
- All living things have basic needs such as air, water, and nutrients. Plants need humans for carbon dioxide, and animals need plants to produce oxygen.
- When we discuss living things and the environment, there are cycles that exist such as the life cycle chain.
- Living organisms are made up of one or more cells. We can develop a deeper understanding of living things by exploring the levels of their organization. We start with the atom and move on to the cells themselves and then to organs, systems, organisms. We will explore how organisms interact with others to form communities and ecosystems later.
- All living things have basic life functions that they are able to perform in order to survive; they are regulation, transport, reproduction, growth and development, respiration, nutrition, and locomotion.
- Biochemistry is the study of chemical substances and the process that occurs in living organisms. The foundation of chemistry is the atom, its structure, and how elements differ.
- Living things need inorganic and organic compounds to survive. The organic compound groups are carbohydrates, lipids, proteins, and nucelic acids.

BRAIN TICKLERS—THE ANSWERS

Set # 1, page 4

1. L
2. N
3. L
4. L
5. N
6. N
7. N
8. L

9. baby/infant → child → teen/adolescent → adult
10. The change occurred due to weathering and erosion. Water and wind acted on it over a long period of time to create the holes and crevices.

Set # 2, page 8

1. air, water, and nutrients (food)
2. d. moving materials
3. c. Air and water are necessary for both plants and animals to survive.
4. fossil
5. a. A nonliving thing that occurs in nature.
6. air and water
7. Everything on the earth exists together to result in a beautiful balance. This statement explains how living things need each other and nonliving things to survive. Animals need oxygen from the plants to breathe; in exchange the animals give carbon dioxide to the plants so they can survive. Nonliving materials like sunlight and water help the living things survive by giving energy from sunlight and enabling them to make food and move materials with the help of water.

Set # 3, page 10

1. homeostasis
2. b. cells → tissues → organs → organ systems → organism

Set # 4, page 15

1. a. growth
2. b. metamorphosis
3. The fecal wastes that animals get rid of are from unused foods that the animal has eaten.
4. Plants
5. Growth and development

6.

a.	5. Movement to get shelter
b.	3. Materials move in organism
c.	2. Respond to stimuli
d.	6. Metamorphosis
e.	1. Breathing
f.	7. New offspring
g.	4. Ingest, digest, egest

7. Answers may vary.
8. make its own food

Set # 5, page 18

1. nucleus
2. b. orbital
3. A, orbital or electron cloud; B, nucleus
4. Elements differ from each other based on the number of protons in the nuclei of their atoms.
5. Matter is anything that has mass and takes up space.

Set # 6, page 23

1. atom
2. c. molecules
3. a. compound
4. periodic table of elements

Set # 7, page 26

1. Inorganic compounds have either carbon or hydrogen atoms or neither, and organic compounds have both carbon and hydrogen.
2. b. The organic compound used for growth and repair is the protein.
3. d. nucleic acids

4. a. carbohydrate
5. Water is made up of the elements hydrogen and oxygen, which make it an inorganic compound.
6. lipids and/or proteins
7. All organic compounds are made up of carbon. It is the foundation of energy.

CHAPTER TWO

Classification

There are so many different types of living organisms on the earth. To help study them, scientists thought of ways to organize all of the living things into groups according to their similarities and differences.

Over time scientists have worked very hard to organize living things into groups so that it is easier for them to be studied. **Classification** is the process of grouping organisms based on their similarities. **Taxonomy** is the study of how living things are classified.

Living things are classified by shared **characteristics** on the cellular and organism level. In classifying organisms, biologists considered details of their internal and external structures. All classification systems are arranged from **general (kingdom)** to **specific (species)**. Classification schemes reflect orderly patterns and observable differences among organisms. Scientists consider an organism's structural features more important for classifying organisms than behavior or general appearance.

The **levels of classification** from largest to smallest are

Kingdom
Phylum
Class
Order
Family
Genus
Species

Here is an example of how an animal would be classified.

Kingdom—Animalia
Phylum—Chordata
Class—Aves
Order—Passeriformes
Family—Corvidae
Genus—*Cyanocitta*
Species—*cristata*
Common name—Blue Jay

You may not realize it, but you use classification in your daily life. Here is an example. If you go into the supermarket to buy something, how do you know where to find the item you're looking for? All of the products in the store are sorted and organized by different characteristics. You are shopping for a Granny Smith apple. How would you find the apple? You would go to the produce section of the store and then look for fruits, apples, and finally the Granny Smith apples.

Carolus Linnaeus developed the system that we use today. Linnaeus placed organisms into different groups based on their observable characteristics. The more classification levels that organisms share, the more characteristics they have in common. Linnaeus gave each organism its own two-part scientific name; he called this system **binomial nomenclature**. These two names are also known as the genus and species.

Remember that the biggest group is the kingdom and the most specific is the species.

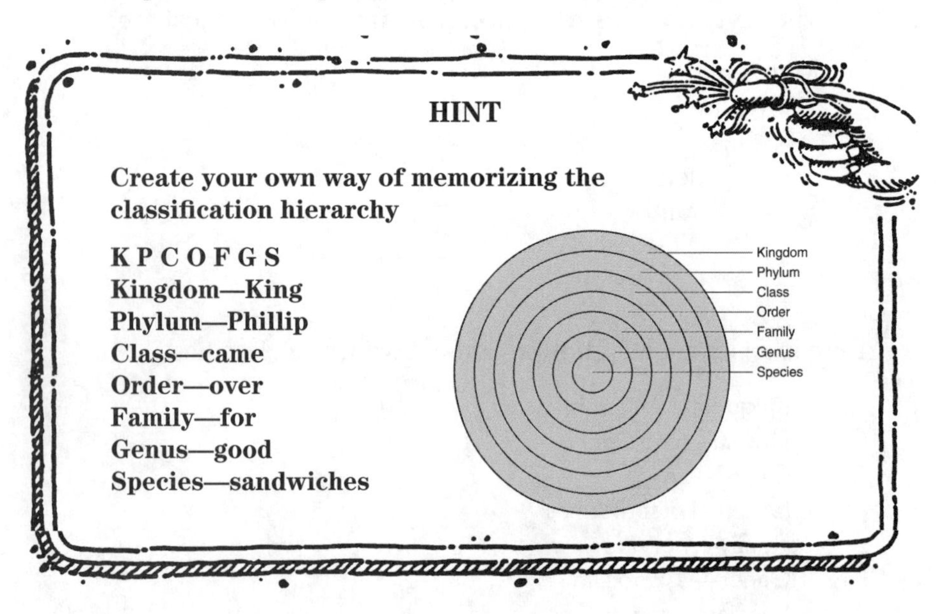

Binomial nomenclature is the formal system of naming specific species by using two names, the genus and the species. This name is often called the scientific name. Look at the following diagram; all of these "cats" belong to the same genus but different species.

genus and species—common name

Felis rufus—bobcat
Felis domesticus—house cat
Felis concolor—cougar
Felis leo—lion

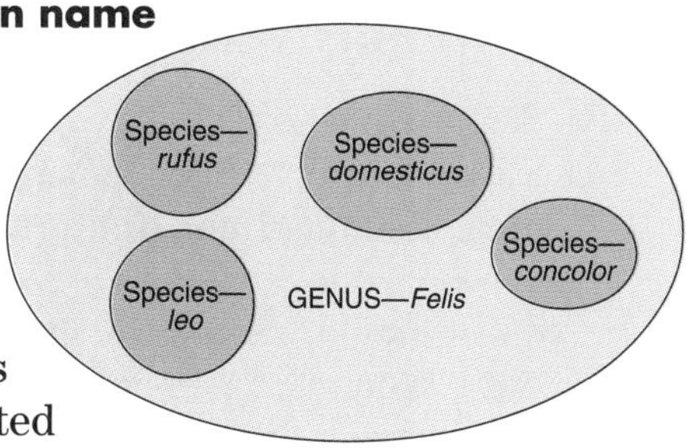

One characteristic that is considered when classifying animals is cell structure. They can be separated into two groups;

Prokaryotic—one cell with no nucleus
Eukaryotic—one cell with nucleus

Other characteristics are the number of cells they are made of, whether they can move, and whether they can make their own food. The five kingdoms are listed below.

Kingdoms

Kingdom	Cell type	# of cells	Mobile	Traits	Examples
Bacteria	Prokaryotic	One cell	Some move	Some can make own food; others can't	E. coli
Protista	Eukaryotic	One cell	Some move	Live in watery environment	Amoeba, euglena, paramecium
Fungi	Eukaryotic	One and many cells	Don't move	Obtain food from others	Mushrooms, mold
Plant	Eukaryotic	Many cells	Don't move	Make own food	Trees, flowers, vegetables
Animalia	Eukaryotic	Many cells	All move	Eat plants and other animals	Human, elephant, mouse, dog, whale

Diversity within a species

Scientists who classify organisms have keys that they use to help them identify the genus and species of an organism. These keys are called the **dichotomous key**. It helps you work through a series of questions and answers. If you answer yes to one question, it tells you the number of the next question to move onto. When you find enough traits in common, it lists the name of the species you are looking at.

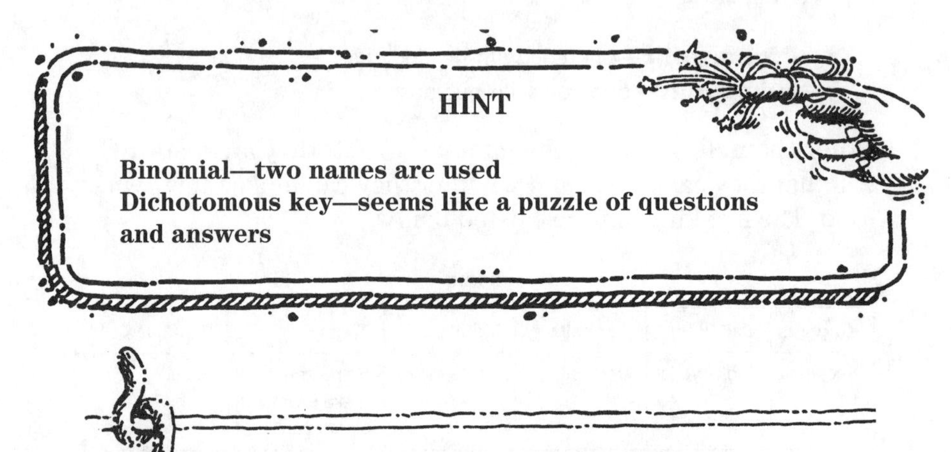

BRAIN TICKLERS
Set # 8

1. Imagine your teacher asked each student in your class to throw one shoe into the middle of the room. Create a list of characteristics that you could use to classify the different types of shoes.
2. The system by which yes and no questions direct you to the genus and species of an organism is called __________.
   a. binomial nomenclature
   b. kingdom, phylum
   c. dichotomous key
   d. classification leveling

3. Which category do the largest number of organism fit into?
4. Which category do the closest related organisms fit into?
5. Why was it necessary to develop a classification system?

(Answers are on page 47.)

PROTISTA AND BACTERIA

There are several microscopic single-celled organisms in the world; some are helpful, others are harmful. We will look at both.

Bacteria

Bacteria can be found in your home, your school, and your body. They are classified by their shape. Bacteria have three different shapes: round (spheres), rods, and spirals.

Cyanobacteria are a type of bacteria that have chlorophyll and are able to make their own food.

Bacteria have many uses; some are used to make cheese or yogurt. **Pasteurization** is a process by which food is heated to a temperature that will kill any harmful bacteria. This is done to milk and some fruit juices.

Other bacteria are used to make **antibiotics** that cure diseases. The word *antibiotic* means "against life" (bacterial). A common antibiotic is penicillin, which was discovered by mistake from a type of mold.

Some viral and bacterial diseases can be prevented by vaccines. A **vaccine** is made from parts of killed viruses and bacteria. For example, when your body is injected with a vaccine, it then learns to recognize the virus and bacteria and fight it. So, if and when your body gets the virus or bacteria, it is able to fight it with white blood cells.

Lyme disease is caused when an infected tick bites a person and infects the blood with the *Borella burgdorferi* bacteria.

The *Mycobacterium tuberculosis* bacterium produces a disease in humans called **tuberculosis** or **TB**. TB is spread by close contact with other people who have the disease and thus the bacteria. TB attacks the lungs and, if not treated, can be fatal. Scientists have been able to develop a vaccine to fight TB. Patients with TB can also be given antibiotics to fight the disease.

Protista

A protist is a simple microscopic single-celled organism that lives in moist or wet surroundings. All protista have a nucleus and are considered **eukaryotes**.

Plant-like protista are known as algae. The **euglena** is a green single-celled organism that has chloroplasts and is able to make its own food.

Animal-like protista are known as protozoa. The **amoeba** is a single-celled organism that has no definite shape. It moves by pushing forward its cytoplasm into portions of its cell membrane. The projections are called **pseudopods**.

The **paramecium** is a single-celled organism that is shaped like a slipper. It moves about with the help of tiny hairs that surround the cell called **cilia**.

FUNGI

Fungi are an important part of nature. Fungi *recycle dead* organic matter into useful nutrients. Many plants, however, are dependent on the help of a fungus to get their own nutrients.

Fungi digest food outside their bodies: They release enzymes into the surrounding environment, breaking down organic matter into a form the fungus can absorb. Examples of fungi are molds and mushrooms.

PLANT

The **Plant** Kingdom consists of all living organisms that are multicellular and are able to make their own food through *photosynthesis.*

Plants can reproduce by one parent (asexual) or two parents (sexual). Asexual plants have offspring that are identical to the parent. They usually drop spores to the ground from the underside of the leaf and a new plant grows. Ferns reproduce this way. Other asexual plants create long stem-like extensions called runners from which new plants grow. Strawberries reproduce this way.

Seed plants are the result of sexual reproduction. They result from pollen in the male part of the plant joining with the ovary in the female part of the plant. The seed is created and will develop into an offspring that can have the traits of both parents. We will go into how the plant reproduces in another chapter.

Plants can be classified by the type of seeds they produce. **Monocots** have one cotyledon. The **cotyledon** is the part of the seed that provides food for the seed. The leaf of a monocot plant has the veins running parallel. Corn is a monocot; you can't split the seed into two even parts because it has one cotyledon.

Dicots have two cotyledons. The seed can be split into two halves. Beans and peanuts are perfect examples of dicots. The dicot's leaves have a main vein in the leaf and then other veins that branch off from the main one.

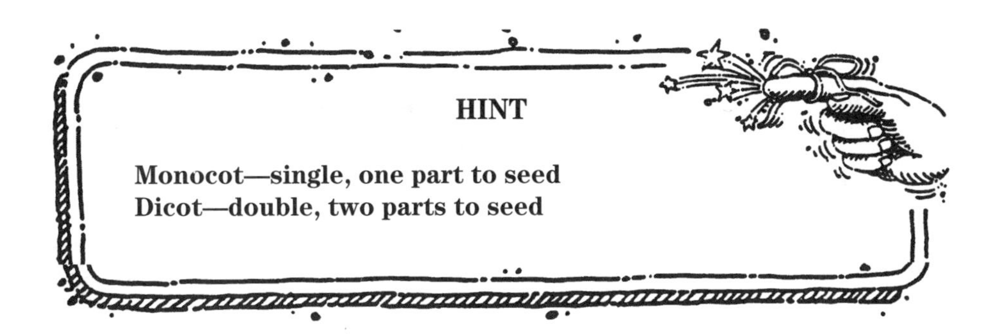

HINT

Monocot—single, one part to seed
Dicot—double, two parts to seed

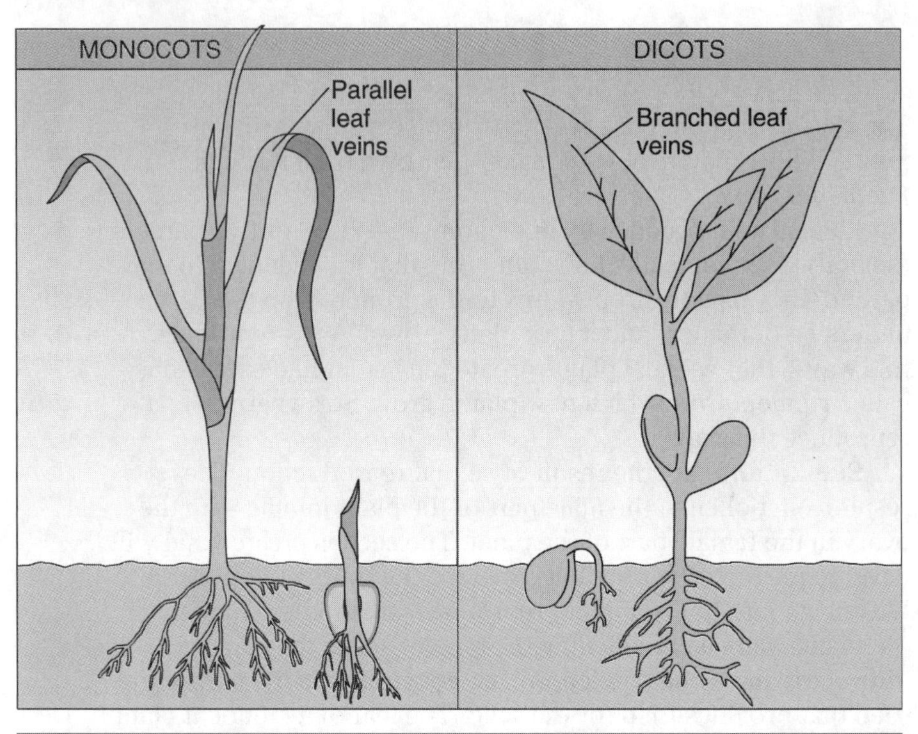

Seeds: contain one cotyledon	Seeds: contain two cotyledons
Flowers: floral parts in multiples of 3	Flowers: floral parts in multiples of 4 or 5
Stems: vascular bundles are scattered	Stems: vascular bundles arranged in a circle
Leaves: long tapering blades with parallel venation	Leaves: broad to narrow leaves with netted venation

BRAIN TICKLERS
Set # 9

1. Which single-celled organism is able to make its own food?
2. Which kingdom recycles dead organic materials?
3. Give two characteristics that separate monocot plants from dicot plants. Use the chart on page 40 to help.
4. Bacteria are classified by their ________.
   a. size
   b. number
   c. color
   d. shape
5. List the five kingdoms in order of smallest to largest.

(Answers are on page 47.)

ANIMALIA

There are millions of species of animals identified and probably many more that still need to be identified. There are nine major phyla of animals.

Here are the basic characteristics of animals:

1. Animals cannot make their own food; some eat plants, some eat animals, and others eat both plants and animals.
2. Animals digest their foods.

3. Animals have the ability to move from place to place. They may move to find food, get protection, or find a mate.
4. Animals have many cells.
5. Animals have a nucleus and organelles in their cells surrounded by a membrane. Their cells are considered eukaryotic.

The animal kingdom is divided into vertebrates and invertebrates.

There are several groups of **invertebrates**. The table shows some of their key features. They *do not have backbones*.

Invertebrate

Group	Examples	Key features
Coelenterates	• Jellyfish • Sea anemone	• Hollow bodied • Mouth is the only body opening • Tentacles
Flatworms	• Liver fluke • Tapeworm	• Flat, thin bodies • Digestive system has one opening
Annelids	• Earthworm • Leech	• Rounded bodies • Bodies made of rings or segments
Molluscs	• Snail • Oyster	• Soft body in three continuous parts with head, foot, and body mass • May have one or two shells
Echinoderms	• Sea urchin • Starfish	• Spiny body in five parts • Central mouth
Arthropods	• Crab • Spider • Fly • Centipede	• Hard body divided into segments • Jointed legs

There are five groups of **vertebrates**. The table shows some of their key features. They *all have backbones.*

Vertebrate

Group	Examples	Key features
Reptiles	• Crocodile • Lizard • Snake	• Cold-blooded • Have lungs • Dry scaly skin • Lay leathery-shelled eggs
Amphibians	• Frog • Newt • Salamander	• Cold-blooded • Adults have lungs; larvae have gills • Moist skin • Lay jelly-coated eggs in water
Fish	• Goldfish • Shark • Cod	• Cold-blooded • Have gills • Wet scales • Lay eggs in water
Birds	• Budgerigar • Sparrow • Ostrich	• Warm-blooded • Have lungs • Feathers on body • Lay eggs with hard shells • Have wings
Mammals	• Rabbit • Kangaroo • Human • Dolphin	• Warm-blooded • Have lungs • Have body hair or fur • Give birth to live young • Produce milk

Fish are chordates that are able to live in the water; they breathe through their gills. The gills on the sides of their heads are where the water gets filtered to take the oxygen out.

Amphibians (Greek—"double life") are vertebrates that spend part of their life in the water and part on land. They have moist skin that is thin without scales. They exchange gas through this skin and the lining of the mouth. Their lungs are very simple sacs. Their body temperatures change with their surroundings. In the cold months, they are inactive and go into hibernation; in the warm weather, they are active. They lay their eggs in the water; they do not have a hard shell. Their offspring are born in a larva form and go through a series of changes over their lifetimes called metamorphosis.

Reptiles are vertebrates. They have scaly skin that keeps their bodies from drying out. Their young do not go through a larva stage like amphibians do; instead, they look like small versions of the adults when they hatch. Reptiles are ectotherms, so they must bask in the sun or find a warm spot to get warm and become active, and they must find shade or a cool spot to cool off. In cold conditions, they become sluggish and don't move around much. In fact, some enter a state of torpor or hibernation if it will be cold for a long time. Over time, reptiles adapted to living on the land. They have claws, and their eggs have a leathery covering. Reptiles lay many eggs in the hopes that some will survive.

Birds belong to the Class Aves. They are characterized by feathers, a beak with no teeth, hard-shelled eggs, a high metabolic rate (which is why they eat constantly), a four-chambered heart, and a lightweight but strong skeleton. Birds are built for flight; they have hollow bones and wings that help them fly.

Some birds can't fly; a penguin is an example of a flightless bird. They lay eggs like reptiles; they have amniotic sacs that hold all of the nutrients the developing embryo will need to survive until it is hatched. Birds provide care for their young after they are born. Many birds move to warmer areas when cold weather moves in; this is called **migration**.

Mammals belong to the Class Mammalia. They are characterized by sweat glands, glands that are able to produce milk, and a highly defined brain. They give birth to live young instead of laying eggs, and they possess specialized teeth. There are over 5,400 species of mammals that range in size from the bumblebee bat (30–40 mm) to the blue whale (33 m).

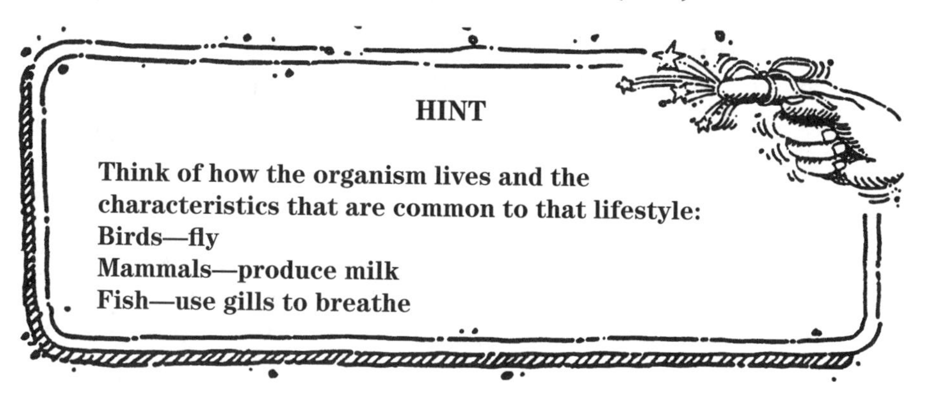

Human classification is

Kingdom: Animalia
Phylum: Chordata
Subphylum: Vertebrata
Class: Mammalia
Order: Primates
Family: Hominidae
Genus: *Homo*
Species: *sapiens*

BRAIN TICKLERS
Set # 10

1. The closest relative to *Canis lupus* is
   a. *Homo sapien*
   b. *Canis familiaris*
   c. *Felis tigris*
   d. *Equus zebra*
2. Prokaryotes are cells with ________.
3. How did mammals get their name?
4. How do birds' hollow bones help them?
5. Which class of animals goes through metamorphosis?

(Answers are on page 47.)

WRAPPING UP

- Classification is the process of grouping organisms based on their similarities. Taxonomy is the study of how living things are classified.
- Organisms are classified into Kingdoms such as Plants or Animalia. They are then divided into more specific groups based on a series of characteristics such as being a vertebrate (having a backbone) or an invertebrate (having no backbone).
- The system of giving a scientific name to an organism is known as binomial nomenclature—organisms are recognized by their genus and species. Species is the most specific of the groupings.
- Unicellular organisms belong to the Kingdom Protist.
- Plants are classified by how they reproduce, the type of seeds they produce, the number of seeds they form, and the structure of their leaves as well as other characteristics.

BRAIN TICKLERS—THE ANSWERS

Set # 8, page 36

1. Answers will vary. Possible list: sneakers—color, size, brand, stripes, laces; sandals—flip-flops, ankle straps; dress shoes—laces, slip ons.
2. c. dichotomous key
3. kingdom
4. species
5. Answers may vary. There were so many animals that scientists needed a way to sort and organize organisms.

Set # 9, page 41

1. Euglena or cyanobacteria
2. Kingdom Fungi
3. Answers will vary. Possible answers include root structure, seed structure, and leaf vein structure.
4. d. shape
5. Protista, Bacteria, Fungi, Plant, Animalia

Set # 10, page 46

1. b. *Canis familiaris*
2. Prokaryotes are cells with no organized nucleus.
3. Mammals have mammary glands that produce milk.
4. Birds' hollow bones keep them lighter for flight.
5. Amphibians—frogs

CHAPTER THREE
Cells

Cells are the smallest individual beings that have certain tasks or jobs that they do on their own or with other cells. Some organisms are made of only one cell; others, like us, are made up of many cells. Some cells live a long time; others live only a short time. As you will learn in this chapter, cells are the foundation of all living things!

The shape and contents of a cell depend on its job. Some different types of cells are red blood cells, plant cells, sex cells, and nerve cells.

CELL THEORY AND MICROSCOPES

Robert Hooke, an English scientist, is responsible for discovering the cell. In 1665 he took a piece of cork and cut a thin slice off. He looked at it under a microscope and noticed that the cork had little chambers or rooms that he called cells. We call rooms in a jail "cells."

Two hundred years later, a German scientist named **Matthias Schleiden** used the microscope to study plants, and he stated that all plants were made up of cells.

Around the same time, another German scientist named **Theodor Schwann** studied animal tissue under the microscope, and he said that all animals are made up of cells.

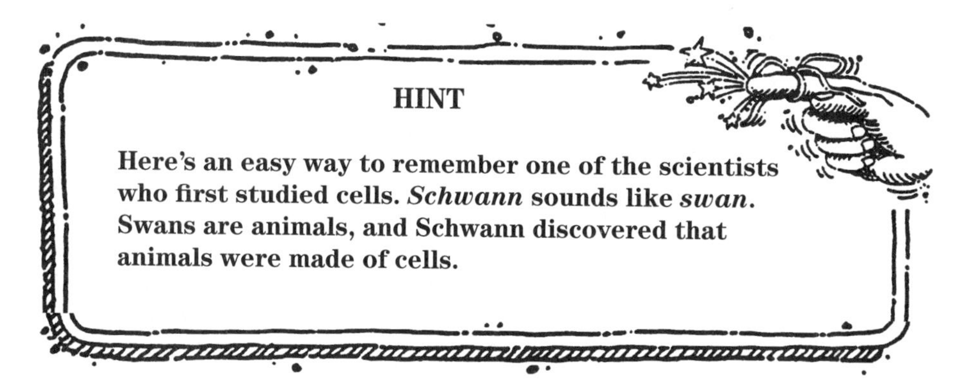

HINT

Here's an easy way to remember one of the scientists who first studied cells. *Schwann* sounds like *swan*. Swans are animals, and Schwann discovered that animals were made of cells.

In the late 1800s another German scientist named **Rudolph Virchow** came up with the **cell theory**. To this day it is the foundation of all cell biology.

The Cell Theory

1. **All organisms are made up of one or more cells.**
2. **Cells are the basic units of structure and function in all organisms.**
3. **All cells come from cells that already exist.**

BRAIN TICKLERS
Set # 11

1. The English scientist named __________ cut a small piece of __________ and named the chambers __________.
2. Schleiden and Schwann discovered that __________ and __________ are made up of cells.
3. An idea that was not part of the cell theory is that __________.
   a. all cells carry on their own life processes
   b. all organisms are made up of one or more cells
   c. all cells are surrounded by a cell wall
   d. new cells arise only from other cells

(Answers are on page 85.)

MICROSCOPES

The **microscope** is an important tool used by scientists. Cells are very small and in order to see a cell, scientists use the microscope.

Compound light microscope

The **compound light microscope** can help you see an object larger and clearer. Depending on the power of the lenses on a microscope, you can see an object anywhere from 100 times, 400 times, or even 1,000 times larger. You would be able to see the blood flowing through a goldfish's tail or the detail in the wing of a fly. By using a microscope, we are able to see how living things are made.

The object you want to examine under the microscope is called a **specimen**. Your specimen must be placed on a slide with a **cover slip** over it. The slide and cover slip can be made of clear glass or plastic. The specimen should be wet to help the light pass through the object so you can see it better. To increase the magnification, you move the **objective lens** to a higher power.

The area that you see through the microscope is called the **field of vision**. The **magnification** of your microscope lens determines how big your field of vision is and how much you can see. The easiest way to see how big an object really is, is by placing a microscopic ruler across the field of view.

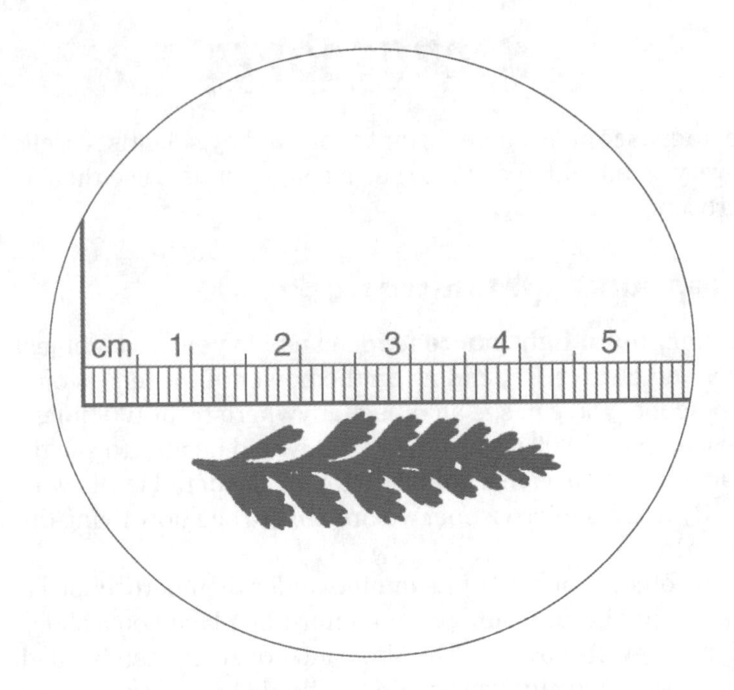

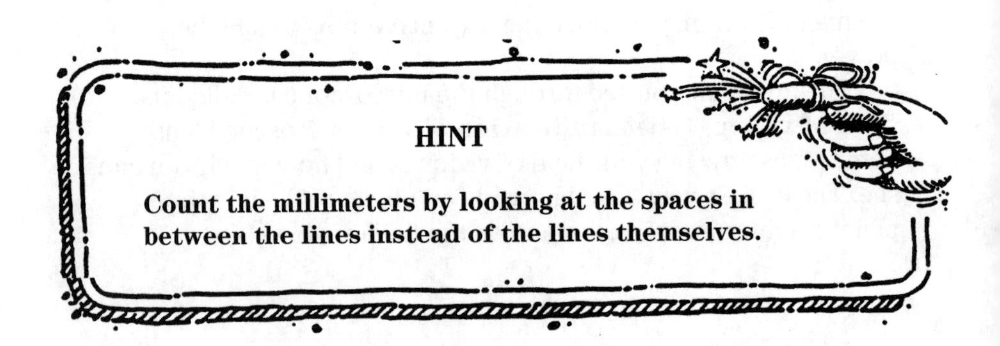

HINT

Count the millimeters by looking at the spaces in between the lines instead of the lines themselves.

The following examples show what the field of vision for a specimen would look like under different magnifications:

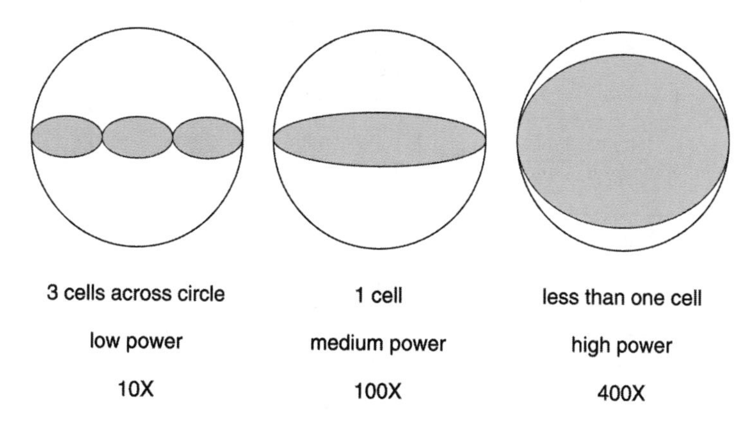

It is important for you to know the parts of the microscope and how it works. Here is a diagram of the microscope with its parts labeled. Think of yourself looking through the **eyepiece** at a mystery object you want to learn more about.

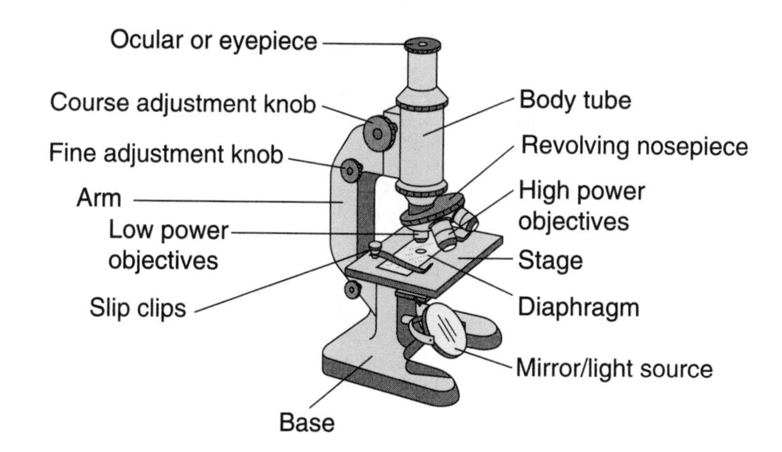

Parts of the Microscope

Structure	Function
Eyepiece – ocular	The place where you look into the scope. The first lens that magnifies the image.
Body tube	Tube of mirrors that connects the eyepiece lens with the objective lenses.
Arm	Joins the lenses to the rest of the microscope. This is where you place one hand when carrying a microscope.
Stage	Just like a play or concert, this is where the star—the slide of your specimen—is placed.
Stage clips	Hold the slide in place.
Diaphragm	An opening that controls the amount of light passing through to the lenses.
Course adjustment knob	Controls how high or low the lens needs to move to focus the object being looked at. This is only used when the objective is on low power.
Fine adjustment knob	Gives a sharper focus of an object being viewed at medium and high objective power.
Light source	In order to look at an object, you need light. Some microscopes use mirrors, others a light.
Base	The bottom of the microscope. It is where your other hand is placed when carrying a microscope.

Stereo microscope

You may use a **stereo microscope** in school. It is used to observe larger objects such as rocks, flowers, or animals. There is an ocular lens for each eye. Because the objects that you look at with this microscope are thick, the light does not pass through the object.

Electron microscope

Objects that are too small to be seen with the compound light microscope can be seen in the very special **electron microscope**. This microscope is very large and it takes a great deal of training to learn how to use it. After preparing the specimen, you can take photographs of the object that are one million times larger. A very detailed photograph is taken of the object under the magnification.

BRAIN TICKLERS
Set # 12

1. How do you increase the magnification of an object while using the compound light microscope?
2. Which microscope can magnify an object up to one million times larger?
3. When you look into a microscope you look through the __________.
4. If you want to look at a large object in more detail you would use a __________ microscope.
5. Which power would allow you to see an object in the most detail
   a. low power
   b. medium power
   c. high power
   d. none of the above
6. The largest field of vision is shown at which power?
   ____

(Answers are on page 85.)

CELL ORGANELLES—STRUCTURE AND FUNCTION

The cell is the foundation of all life. Every cell has a different job depending on where a cell is located. Each cell is made up of different parts called **organelles**. Each organelle has a specific job to do and with each organelle doing its own job, the organelles work together like a machine, a school, or a town to get their job done. Let's learn about a cell and its organelles. If we compare it to a working town, it is easier for us to remember and understand the different jobs and how the organelles work together.

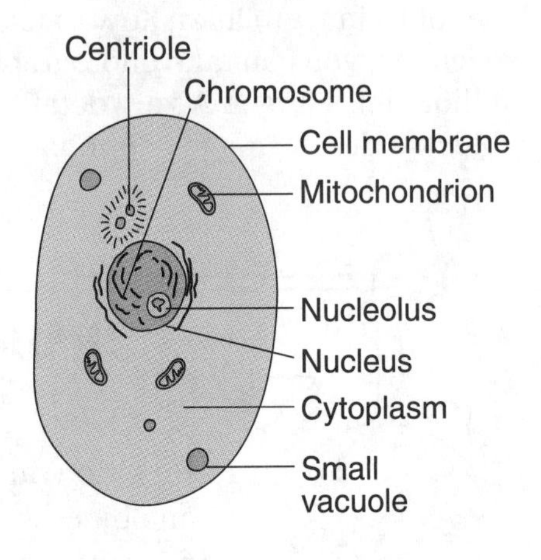

The Cell as a Community

Organelle	Function	Community Comparison
Cell membrane	Regulates the transport of materials into and out of the cell	Border patrol that surrounds a town
Cytoplasm	Fluid-like environment in which organelles are suspended	Air that everything moves around in
Nucleus	Control center of the cell; directs all activities of the cell; Genetic Information—DNA/ chromosomes—found here	The town hall; rules are made here and supervised
Nucleolus	Makes (synthesis) ribosomes	The mayor who makes the rules
Ribosomes	Site of protein synthesis; may be free in the cytoplasm or attached to membranes	Workers who come from the town hall and move around town
Endoplasmic recticulum	Interconnection channels associated with storage and transport of material through the cell	Roads all around the town

Mitochondria	The "Power House" of the cell; site of cellular respiration	Power plant that makes energy for the town
Golgi Bodies	Synthesis, packaging, and getting rid of cellular wastes	The garbage men
Lysosomes	"Suicide Sac" contains digestive enzymes; if it breaks, it will kill the cell	Clean up crew—they get rid of garbage and waste
Vacuole	Space that holds water and other materials	Storage warehouse or reservoir
Centriole	Cylinder-shaped structure that helps in cell division in animal cells	Helper for organizing reproduction
Chloroplast	Contains the green pigments in the plant cell that are needed for photosynthesis	Only in plants; the chef who makes the food
Cell Wall	Non-living surrounding of the plant cell that is responsible for support, structure, and protecting the cell	Wall that surrounds the town and protects it

Cytoplasm is the water-like fluid that all of the organelles **float** around in.

The **nucleus** is the *control center* of the cell. It directs the other organelles to do their jobs, whether it is to fight infection, carry nutrients, or heal cuts, among other activities. All genetic information is found in the nucleus on X-shaped bodies called chromosomes. The chemical name for this information is **DNA**, deoxyribonucleic acid. The DNA doesn't leave the nucleus, but it sends the information to the rest of the cell by giving the messages to another chemical called **RNA**, ribonucleic acid. DNA and RNA are similar, but RNA is able to leave the nucleus.

The **cell membrane**, which *surrounds* the cell, is an amazing organelle. It is made up of fats (lipids) and proteins. The fats act like a fence and keep everything out, but the proteins, which are throughout the cell membrane, act like a gate and allow certain things to pass in and out of the cell.

The **mitochondria**, the bean-shaped organelle with a wavy center, *creates energy* for the cell. It does this by taking a broken up piece of sugar from the cytoplasm and passing it through its folded insides. This movement back and forth builds up energy, just like the energy you make when you rub your feet on a carpet to get a shock. The more mitochondria a cell has, the more active it is.

The **endoplasmic reticulum** is the *highway system* of the cell. The endoplasmic reticulum is often called ER. The ER has lots of folds and runs throughout the cytoplasm of the cell. The ER is lined with little ribosomes. If the ER has lots of ribosomes on it, the cell is active because it is moving a lot of material around.

The **golgi bodies** look like stacked pancakes. The golgi bodies work like a *garbage collector*. They package up the waste and take it to the cell membrane for disposal.

Waste products from the cell are held in sacs called **lysosomes**. The lysosome holds chemicals in it that break down waste, worn out cell parts, and pieces of food. The chemicals in the lysosomes are very acidic. If they break, it will kill the cell, which is why the lysosome is also known as a suicide sac.

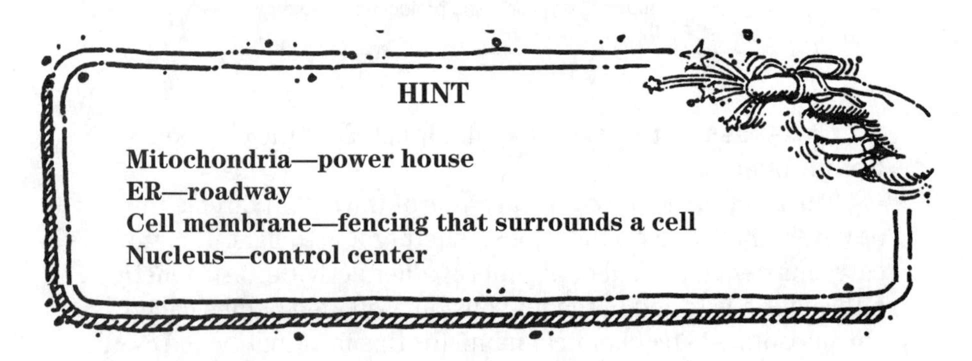

HINT

Mitochondria—power house
ER—roadway
Cell membrane—fencing that surrounds a cell
Nucleus—control center

Plant cells have **cell walls**. The cell wall is on the outside of the cell membrane and is there to give the cell its shape. It also helps to *protect* the cell.

Plant and animal cells have **vacuoles**. They *store* water, food, and other materials needed by the cell. In plant cells, the vacuoles can become very big. The animal cell can have several small vacuoles.

Plant cells are able to make their own food; this is thanks to the **chloroplasts**. The chloroplasts have stacks in them that turn toward sunlight and absorb the energy from the sun to make food. The light energy is transformed into chemical energy in the form of sugar. Animals get their energy from the sugar *produced* by the plant's chloroplasts. The more chloroplasts found in a plant cell, the more energy/sugar it is making.

Cyclosis is how materials *move* around the cell. Imagine being in a round pool. Everyone else in the pool with you moves in the same direction, making the water act like a whirlpool. That is how materials move in the cell; we call this process cyclosis. It sounds like cyclone.

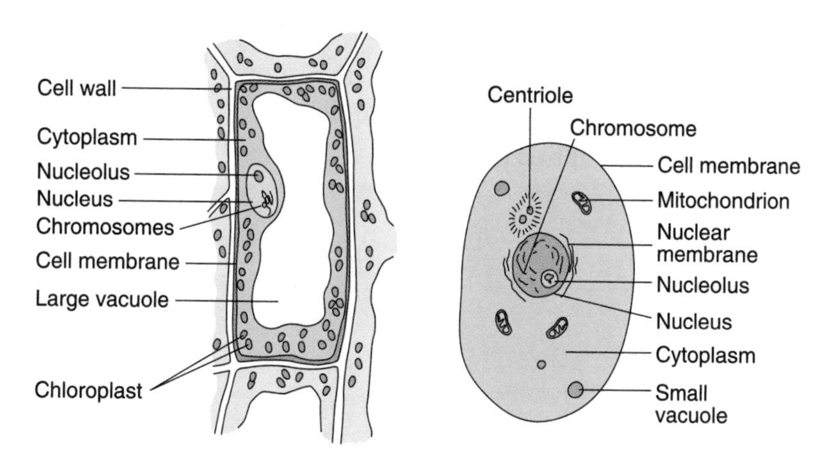

Typical Plant Cell

Typical Animal Cell

Animal Cell	Plant Cell
No chloroplast	Has chloroplast
No cell wall	Has cell wall
Centrioles	No centrioles
Small vacuoles	Large vacuoles

BRAIN TICKLERS
Set # 13

1. Which organelle is in "control"?
2. The highway of the cell is called the ________.
3. Cell structures are also known as ________.
4. What two organelles do plant cells have that animal cells don't have?
5. What organelle is responsible for creating energy for the cell?
6. Label the parts of this plant cell.

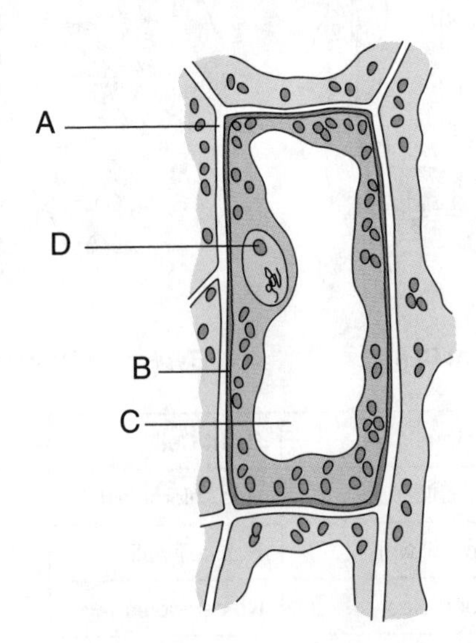

Challenge: Compare the organelles of a cell with the organization of a school building.

(Answers are on page 85.)

CELL MEMBRANE

Each part of the cell has a specific function. The cell membrane is the part of the cell that is responsible for it keeping its shape and for controlling what goes into the cell and what comes out of the cell. The cell membrane acts like a bag holding all the items inside of it. Some people think of it as a fence surrounding a property. When we talk about cell transport, we don't mean how the cell moves around because many cells do not move from one place to the other. Cell transport is how materials move into and out of the cell via the cell membrane. For anything to go into or out of the cell, it has to go through the cell membrane.

The cell membrane is like a **double-layered fence** with spaces scattered throughout. These spaces act like a gate in a fence where certain materials can be moved in and out. The two layers of the cell membrane are made up of little bits of fat, called *lipids*, and larger pieces, called *protein molecules*. The protein molecules act like the gates and the lipids act like the fence. Check out this diagram of the cell membrane.

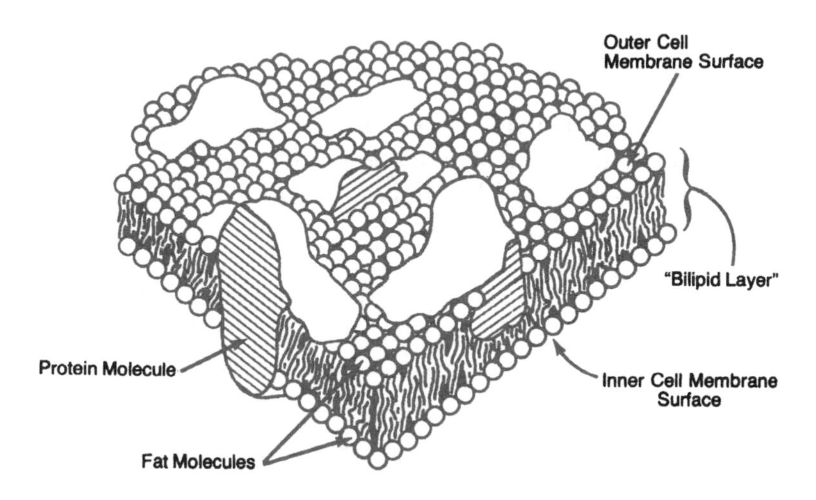

The cell membrane is said to be **semipermeable**. Semipermeable means that some substance can pass through the membrane but that others cannot. The cell determines if an object is too large, if the object is not healthy for the cell, or if the object isn't needed at the time. If the object is judged to be harmful, the cell will not let it in.

The two lipid layers of the cell shift back and forth on top of each other. When the proteins line up together, they create paths/tunnels from the outside of the cell membrane to the inside of the cell. This is the path that materials will be able to pass through. This movement is called the **fluid mosaic model**. (A *mosaic* is a picture made up of little pieces.)

The outside of the cell membrane is **hydrophilic**, which means that it likes water; the inside is **hydrophobic**, which means it doesn't like water. This makes the cell membrane a watertight barrier.

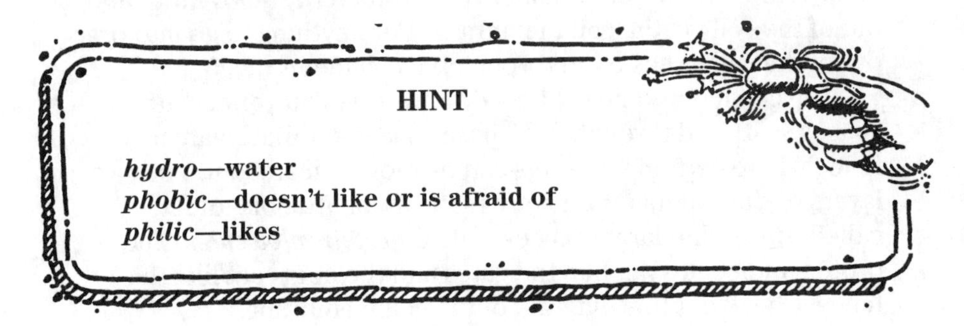

Diffusion

If you stand in the corner of your classroom and spray perfume, do the students in the far corner smell it right away? No, it takes time for them to smell it. What happens is the little particles of fragrance from the perfume spread across the room using its own energy and continue to move as it maintains equilibrium. Where you pressed the perfume would be the area of high concentration; the rest of the room would be low concentration. When the molecules are spread evenly throughout a space, it is at **equilibrium**.

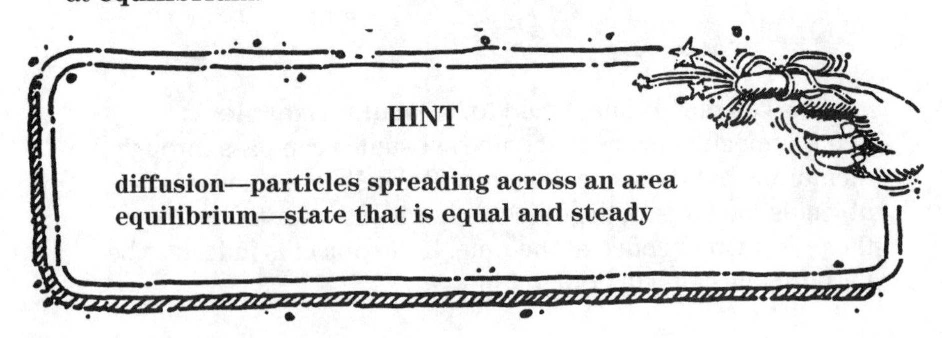

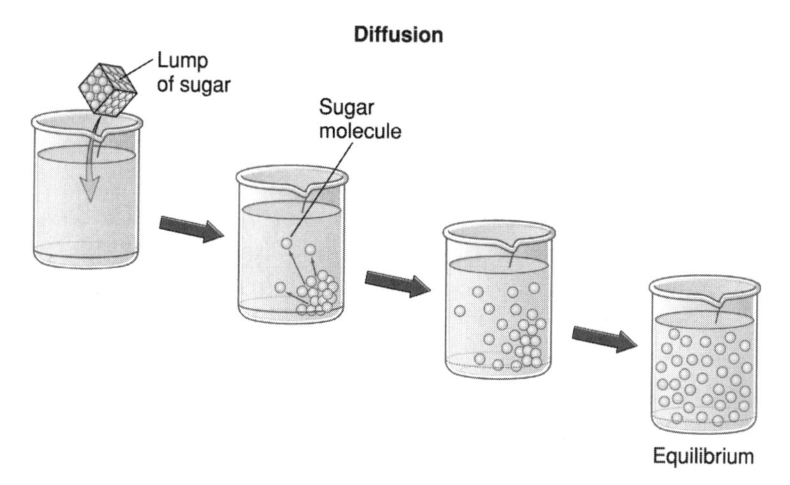

When very small molecules need to move across the cell membrane, we call this diffusion. This occurs only when there is an imbalance of molecules inside and outside of the cell. Molecules in greater concentration will move to areas of lesser concentration as shown in the diagram.

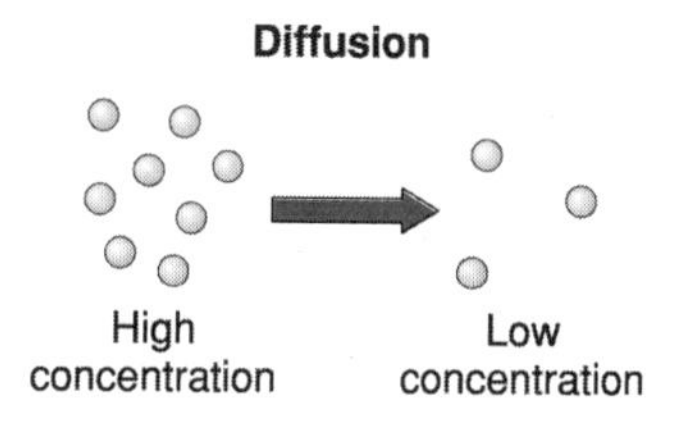

Osmosis

Whenever *water* moves/diffuses through a membrane we call it **osmosis**. Osmosis is the process by which water moves through the semipermeable membrane. Remember that semipermeable means that some things move across.

Active and passive transport

Many things that we experience every day act the same as they do inside of living things. Think about a cart sitting on top of a hill. If the wind blows or someone pushes the cart, it will roll down the hill with no extra effort or work. When an object moves easily from one place to another, we say that it goes with the flow. This is an example of passive transport. No energy is used, and the object moves with ease.

Passive transport occurs when materials move across the cell's membrane and *don't require energy*. They go from an area of high concentration to low concentration. Think of rolling down the hill on a bike: You don't have to peddle or work to get the bike to move.

Active transport occurs when *energy is needed* for the materials to move across the membrane. They go from an area of low concentration to high concentration. Protein molecules move across the membrane via active transport. This would be the same as being at the bottom of the hill and having to peddle or push the bike to get to the top.

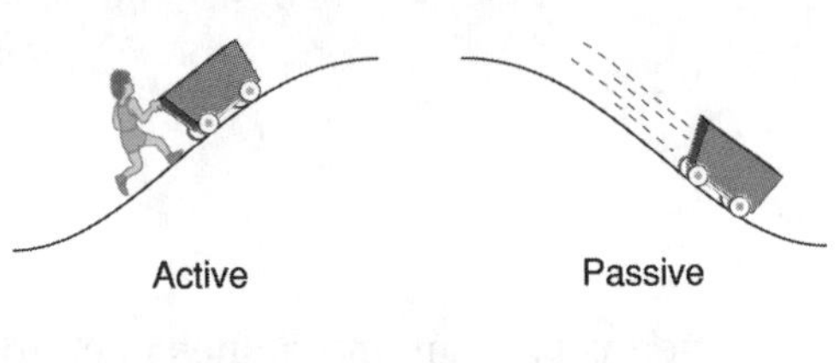

Some substances are too big to pass across the cell membrane so the cell folds over the particles and brings them into the cell. This process is called **endocytosis**. When the cell needs to get rid of wastes, it goes through a process called **exocytosis**. The ball of waste moves to the cell membrane and attaches to it; the cell membrane opens to the outside of the cell releasing the waste to the outside.

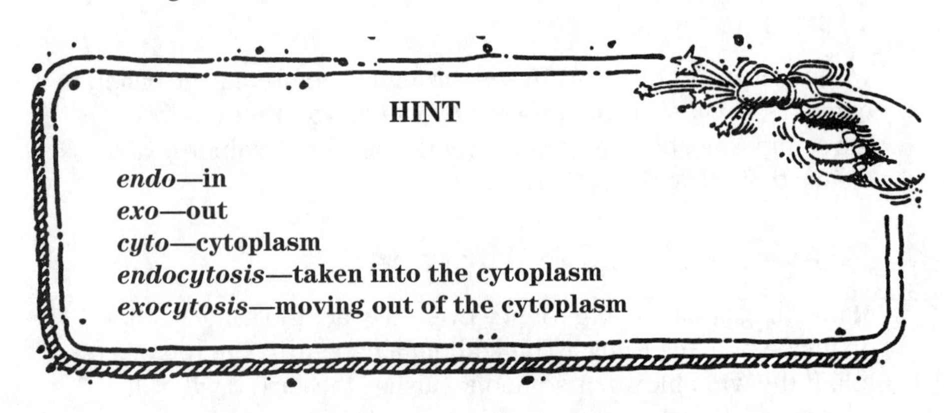

HINT

***endo*—in**
***exo*—out**
***cyto*—cytoplasm**
***endocytosis*—taken into the cytoplasm**
***exocytosis*—moving out of the cytoplasm**

Some cells take in large materials by a process called **phagocytosis**. The cell's cytoplasm pushes into projections called **pseudopods**. These pseudopods are also known as false feet. The fluid moves into the projections, wraps around the food or object, and **engulfs** it.

Cells can take in tiny particles by a process called **pinocytosis**.

BRAIN TICKLERS
Set # 14

1. Which cell structure controls the movement of material into and out of the cell?
   a. nucleus
   b. cytoplasm
   c. cell membrane
   d. vacuole
2. The cell membrane is made up of ________.
   a. lipids and proteins
   b. glucose and lipids
   c. proteins and water
   d. glucose and water
3. Semipermeable means that the cell membrane ________.
   a. allows all materials to pass through it
   b. allows no materials to pass through it
   c. is selective in what passes through it
   d. turns on and off at times
4. Active transport is when ________.
   a. energy is needed for the materials to move across the membrane
   b. water passes across the membrane
   c. sugar moves out of the plant
   d. materials move across the cell membrane and don't require energy
5. When salt is outside a cell, the water in the cell will move toward the salt. This is an example of ________.
   a. diffusion
   b. osmosis
   c. endocytosis
   d. exocytosis

6. When the molecules are spread evenly throughout a space it is at
   a. diffusion
   b. osmosis
   c. equilibrium
   d. respiration
7. The movement of tiny molecules across a space is called __________.
   a. diffusion
   b. osmosis
   c. equilibrium
   d. respiration
8. Pseudopods are also called __________.
   a. false arms
   b. false feet
   c. real feet
   d. real arms
9. Endocytosis and exocytosis are terms to explain
   a. how materials move into and out of the cell
   b. how materials move into and out of the animal
   c. how materials move into and out of the plant
   d. how materials move into and out of the air
10. When objects move from an area of high concentration to an area of low concentration, they __________.
    a. move with ease
    b. don't move at all
    c. move backward
    d. move in circles

(Answers are on page 86.)

CELLULAR ENERGY

Just as a car needs energy to move, cells need energy, too. Animals take in food so that they can get energy. The food is broken down as the animal eats it, and it becomes a nutrient. One nutrient is a kind of sugar called glucose. Glucose and oxygen go into each cell and are converted into energy for the cell and animal to use. The waste of **cellular respiration** in the animal is carbon dioxide, which is exhaled.

Energy produced during cellular respiration is stored in a molecule with a very long name, adenosine triphosphate, called **ATP** for short! ATP is a chemical molecule that has three **phosphate energy** groups. We can represent the molecule as A-P-P-P. Each time a phosphate group breaks off the main molecule, energy is given off and used to do work in the cell.

There are two stages of **cellular respiration**:

> **Glycolysis** occurs in the cytoplasm and is the first stage of cellular respiration. The word *glycolysis* means "to cut up glucose."

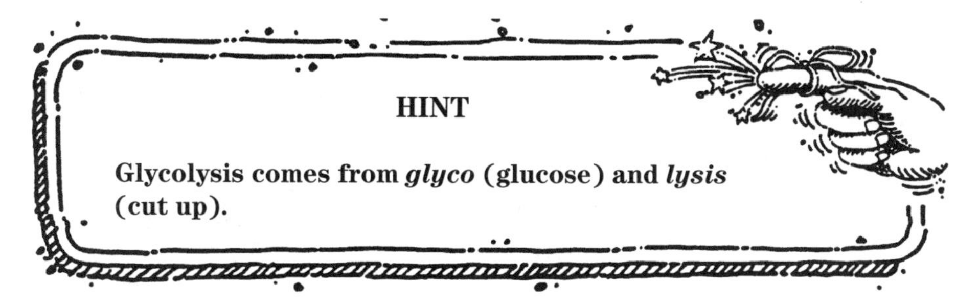

> Glycolysis happens in the cytoplasm of the cell. Glucose enters the cell through the cell membrane and is broken down into smaller molecules; this causes a small amount of energy (ATP) to be released.

The second phase of cellular respiration happens in the **mitochondria** when oxygen is present. The small molecules of three carbon sugars move into the mitochondria along with oxygen; they bounce around and break into even smaller molecules. As they break into smaller molecules, they produce large amounts of energy ATP for the cell to use. The left over parts of the glucose molecule recombine to form carbon dioxide and water.

$$C_6H_{12}O_6 + 6\,O_2 \rightarrow 6\,CO_2 + 6\,H_2O + \text{ATP}$$

glucose + oxygen → carbon dioxide + water + energy

When oxygen is not present, the process of **fermentation** in cells creates energy. Yeast is a single-celled organism that goes through fermentation. The end product of yeast fermentation is alcohol. There are other single-celled organisms that survive without oxygen, and they also go through fermentation.

Lactic acid fermentation happens in the muscles of humans when there is not enough oxygen for cellular respiration to happen properly. When a person plays a sport or exercises without stretching and warming up properly, they can get cramps in their muscles. These cramps come from the pieces of glucose not being able to break down properly. They become lactic acid and stay in the muscles, making them sore.

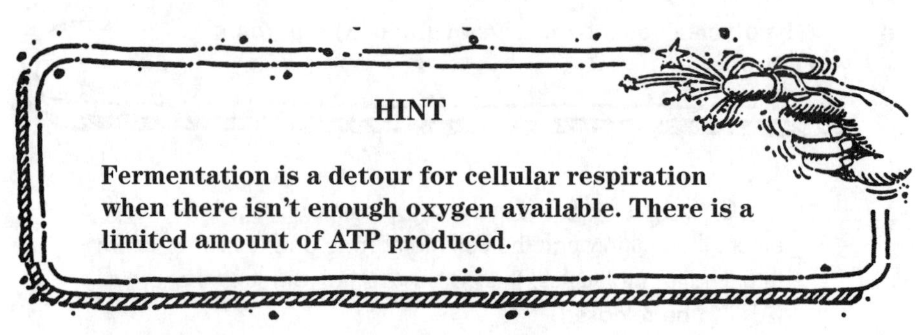

HINT

Fermentation is a detour for cellular respiration when there isn't enough oxygen available. There is a limited amount of ATP produced.

Let's shift gears and talk about how photosynthesis occurs in the plant. Plants are essential for life on earth. Plants take the energy from the sun and converts it into a form that animals can take in and use as energy. Plants absorb energy given off by the sun, take in water and carbon dioxide, and make glucose.

There are two stages of photosynthesis.

Stage 1—Sunlight is taken into the plant to make food. The leaves of the plant act like a suncatcher. The sun shines on the surface of the leaf, and the plant absorbs the sunlight and uses that energy to convert water and carbon dioxide to produce glucose. As part of the process, oxygen and water are given off by the plant. This happens in two stages in the chloroplast.

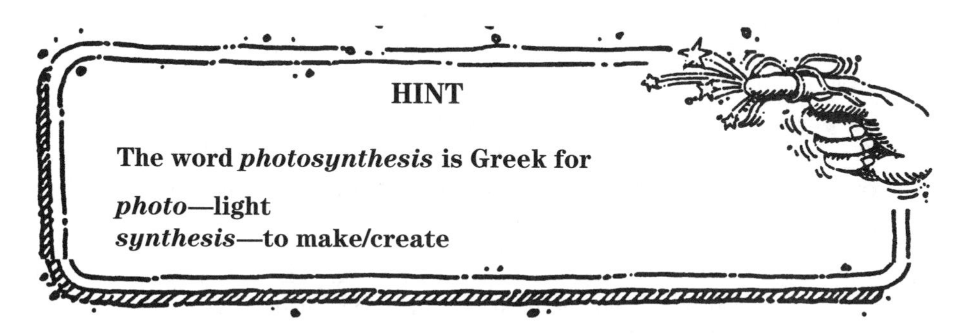

HINT

The word *photosynthesis* is Greek for

***photo*—light**
***synthesis*—to make/create**

Stage 2—Chloroplasts, green organelles found inside plant cells, absorb the light. The green pigment comes from chlorophyll. Plant cells make their own food by using energy captured from the sun in Stage 1:

$$H_2O + CO_2 \xrightarrow{\text{sunlight}} C_6H_{12}O_6 + O_2$$

water + carbon dioxide $\xrightarrow{\text{sunlight}}$ glucose + oxygen

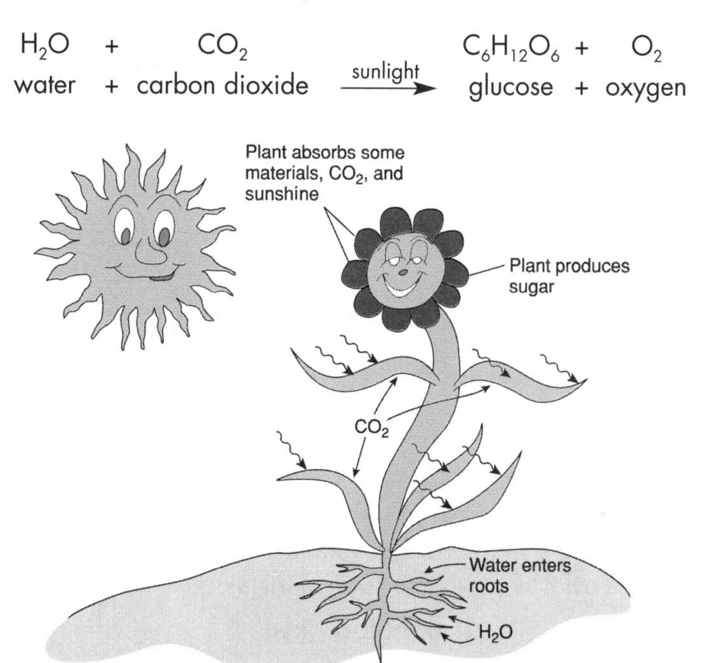

Roots absorb water from the ground and bring water up to the leaf where it is combined with carbon dioxide that is brought in through the leaf. Leaves take in carbon dioxide through specialized openings.

An **autotroph** is an organism that can make its own food. Green plants are autotrophs.

The opposite of an autotroph is a **heterotroph**. Heterotrophs must get their food from other organisms. Animals are heterotrophs because they are not able to make their own food and must take in nutrients to get energy to survive.

Raw materials are items readily found on the earth. The raw materials for cellular respiration in animal cells are oxygen and glucose. The raw materials for photosynthesis are carbon dioxide and water.

Photosynthesis = Chloroplast	Respiration = Mitochondria
Occurs in the presence of light (and chlorophyll) in cells.	Occurs at all times in cells.
Requires energy (light) to make sugar (glucose).	Releases energy from sugar.
Complex substances (sugar) are formed from simpler ones.	Complex substances (sugar) are broken down into simpler ones.
Carbon dioxide and water are the raw materials.	Carbon dioxide and water are the waste products.
Oxygen is given out.	Oxygen is taken in.

BRAIN TICKLERS
Set # 15

1. Which organelle must be in a leaf cell for photosynthesis to occur?
   a. nucleus
   b. cytoplasm
   c. mitochondria
   d. chloroplast

2. Which organelle is only found in plant cells?
   a. cytoplasm
   b. cell wall
   c. cell membrane
   d. nucleus
3. The raw materials needed by the plant for photosynthesis are __________.
   a. carbon dioxide and water
   b. oxygen and water
   c. oxygen and glucose
   d. glucose and carbon dioxide
4. Cellular energy is also known as __________.
   a. sugar
   b. ATP
   c. DNA
   d. ESP
5. The first phase of cellular respiration is glycolysis; this occurs in the __________ of the cell.
   a. mitochondria
   b. chloroplast
   c. cytoplasm
   d. nucleus
6. The largest amount of ATP is produced when cellular respiration takes place in the __________ of the cell.
   a. mitochondria
   b. chloroplast
   c. cytoplasm
   d. nucleus
7. Glucose is broken up during __________.
   a. photosynthesis
   b. active transport
   c. glycolysis
   d. passive transport

8. Water enters the plant for photosynthesis through the __________.
   a. leaves
   b. osmosis
   c. diffusion
   d. roots
9. When there is not enough oxygen present the human muscle will produce __________.
10. List three things to compare animal and plant cellular respiration.

Challenge:
Fill in the arrows with either "oxygen" or "carbon dioxide."

(Answers are on page 86.)

CELLULAR REPRODUCTION

Everything starts from a single cell. One cell may grow in size and exist as one cell for its entire life. Some single-celled organisms called protists include paramecium and amoebas. Multicellular organisms also start as one cell, but their cells divide over and over to grow into a human, horse, ladybug, whale, or tree. This is what happens in all multicelled organisms such as plants and animals.

Reproduction is the process by which an organism produces others of the same kind. Cells are constantly reproducing themselves. When you fall and cut your leg, the skin cells along your cut reproduce and divide to make new cells to close up and heal your cut.

Cell cycle

All cells go though an orderly sequence of changes during their lifetimes called the cell cycle. Most of the cell cycle is spent growing, developing, and preparing for cell division. The cell divides into two new cells called daughter cells. Each daughter cell then goes through the cell cycle again. The cell cycle consists of three phases: interphase, mitosis, and cytokinesis.

Interphase

The cell cycle begins with **interphase**. During interphase the cell goes through rapid growth, **DNA synthesis**, and preparation for division. When the cell grows to its full size, it produces the new cellular organelles it needs such as ribosomes, centrioles, mitochondria, and, if a plant cell, chloroplasts. DNA synthesis occurs during interphase when the cell makes an exact copy of its DNA (this is a process called replication). DNA is usually found as chromatin when the cell is in a dormant (resting) stage and looks granular. Once the DNA has replicated, the cell is ready to go through cell division.

Mitosis

When the cell is ready to divide, it goes through a sequence of changes called **mitosis**. Mitosis occurs when the nucleus divides into two new cells. During mitosis the DNA is split into each of the two new "daughter" cells to make an exact copy of the parent. Mitosis can be broken down into four stages: prophase, metaphase, anaphase, and telophase.

Prophase is the part of mitosis where the cell prepares to split. The chromatin in the nucleus condenses to form double rods called chromosomes. Centrioles move to opposite sides of the cell. Spindle fibers form and attach themselves to the middle of the chromosomes as they stretch across the cell and the nuclear envelope breaks down.

Metaphase is the next phase of mitosis when all the double-stranded chromosomes line up along the middle of the cell in pairs getting ready to split into two. The centromeres attach to the spindle fibers, and the centrioles are positioned at each end of the cell. The ends of the cell are called poles. Once the chromosomes and centrioles are in position, mitosis moves to the next phase.

Anaphase is the phase when the genetic material pulls apart. The centromeres divide, and the two strands of each chromosome pair separate and move toward the centrioles at each pole.

Telophase is the last phase of mitosis when the chromosomes return to a granular appearance as chromatin and the new nuclear envelope forms to show two new daughter cell nuclei.

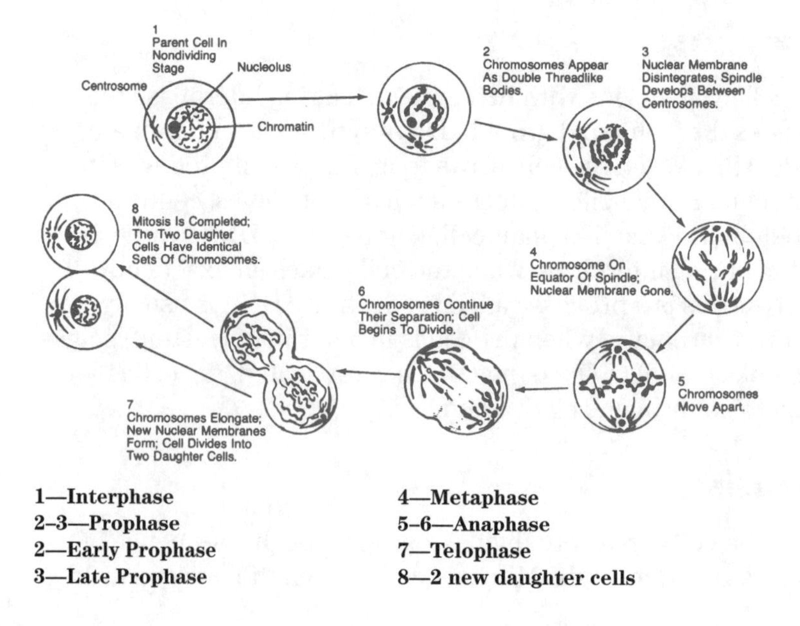

1—Interphase
2–3—Prophase
2—Early Prophase
3—Late Prophase
4—Metaphase
5–6—Anaphase
7—Telophase
8—2 new daughter cells

Cytokinesis

Now that the cell's nuclear material has divided into two exact copies of the original nucleus, the cell's cytoplasm is ready to divide and *separate* the two new cells. This process is called **cytokinesis**. Cytokinesis occurs when the cell membrane pinches in around the middle of the cell and splits the cell into two. In animal cells, during cytokinesis, the cell membrane squeezes in the middle and pinches off to form two new daughter cells that are identical to the parent cell.

Cytokinesis in plant cells is different because the cell wall cannot squeeze in like the cell membrane. A cell plate forms across the middle of the plant cell to form two new daughter cells.

HINT

You can remember the order of the phases of mitosis if you keep in mind IPMAT, or I Poured My Aunt Tea.

interphase—intermediate/ in between
prophase—prepare
metaphase—middle
anaphase—apart
telophase—two new cells

BRAIN TICKLERS
Set # 16

1. Most of the cell cycle is spent ________.
   a. shrinking
   b. getting rid of waste
   c. growing
   d. dividing
2. The cell cycle consists of three major parts: ________, ________, and ________.
3. Mitosis is a process by which the cell goes through a series of changes to produce ________ daughter cells.
   a. two different
   b. four identical
   c. three different
   d. two identical
4. The process by which the cell pinches or splits into two new independent cells is called ________.
5. The centrioles have ________ attached to them to help in cell division.
   a. spindle fibers
   b. chromosomes
   c. centromeres
   d. cell membranes
6. List the five phases of mitosis in order: ________, ________, ________, ________, ________.
7. In which phase do the chromosomes line up along the middle of the cell?
   a. anaphase
   b. prophase
   c. metaphase
   d. telophase
8. Describe what happens to the cell's genetic material during mitosis.

9. The nuclear membrane disappears during __________.
10. In which phase do the chromosomes pull apart?
    a. anaphase
    b. prophase
    c. metaphase
    d. telophase

(Answers are on page 87.)

CHROMOSOMES

Chromatin—DNA during interphase looks like a bowl of rice (granular)

Chromatid—individual thread of a chromosome

Centromere—holds the two chromatids together, where the spindle fibers connect to the chromosome

Chromosome—chromatid and centromere together

Replication of DNA

Replication of DNA occurs in the nucleus during interphase when the genetic material is in its chromatin appearance. DNA looks like a ladder. When it replicates, the ladder unzips and the matching amino acids link up to create copies of the original DNA. DNA is like a puzzle that has a certain code. The sequence of the amino acids makes up the code, and when the DNA replicates, it has the same code as the original.

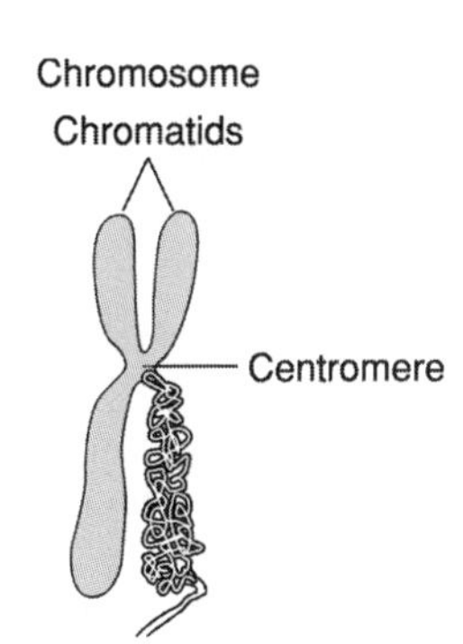

Mitosis in plant and animal cells

Reproduction of body cells in plants and animals happens by *mitosis*. Remember that when mitosis occurs the nucleus is duplicated and then splits into *two new cells*. The original cell no longer exists, but the new cells are copies of the original.

An organism can go through two types of reproduction: asexual or sexual.

Asexual reproduction

In **asexual reproduction**, new offspring are formed from one parent. When an organism goes through asexual reproduction, it goes through mitosis. The DNA in the offspring is identical to that of the parent.

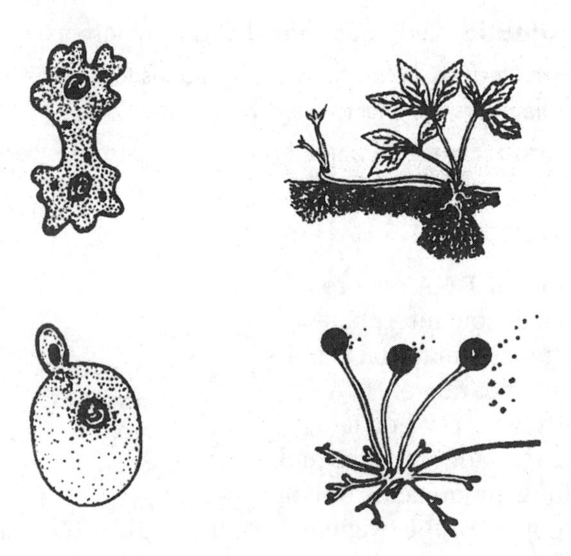

Sexual reproduction

In **sexual reproduction**, a new organism is produced when sex cells from two organisms are joined together. Sex cells contain half of the DNA (genetic material) of the parents' cell. Humans are produced by sexual reproduction. Sperm from the male and an egg from the female combine their DNA to form a new organism. When the egg and sperm join together, it is called **fertilization**. The newly fertilized egg is now called a **zygote**.

sperm	+	egg	=	zygote
23 chromosomes	+	23 chromosomes	=	46 chromosomes

The new organism that results from sexual reproduction is called an **offspring**. The offspring is genetically different from the parents. The offspring has genetic information from both parents.

Meiosis

The sex cells go through two sets of cell division before they are ready for fertilization. **Meiosis** is the process by which the *sex cell* reproduces and divides to prepare itself for fertilization. The phases of meiosis are very similar to those in mitosis but when the second set of cell division occurs, called meiosis 2, the DNA material is not replicated. The phases of meiosis have the same names as those of mitosis, and the same action occurs in each phase.

Meiosis 1	**Meiosis 2**
Interphase	No replication of DNA
Prophase	Interphase
Metaphase	Prophase
Anaphase	Metaphase
Telophase	Anaphase
	Telophase

Meiosis results in four new sex cells with half of the "haploid" number of chromosomes as the parent. The full number of chromosomes in all body cells is called the diploid number.

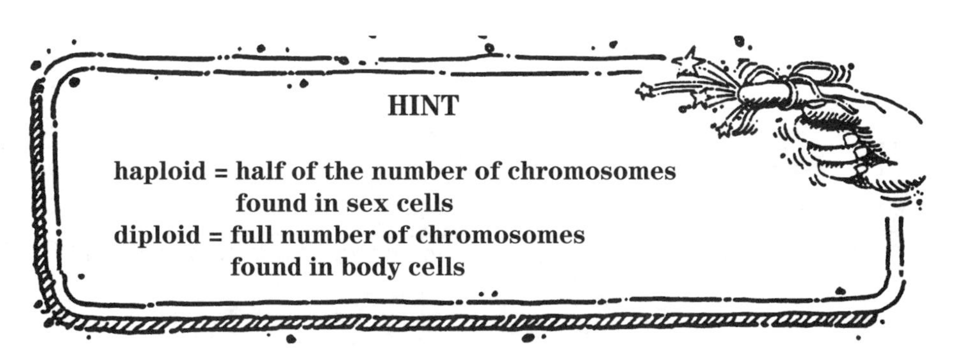

HINT

haploid = half of the number of chromosomes found in sex cells
diploid = full number of chromosomes found in body cells

Chart and pictures of organisms listing the number of chromosomes in body cells and sex cells:

Organism	Number of Chromosomes in Body Cells	Number of Chromosomes in Sex Cells	Number of Chromosomes in Offspring
Human	46	23	46
Fruit fly	8	4	8
Camel	80	40	80

MEIOSIS

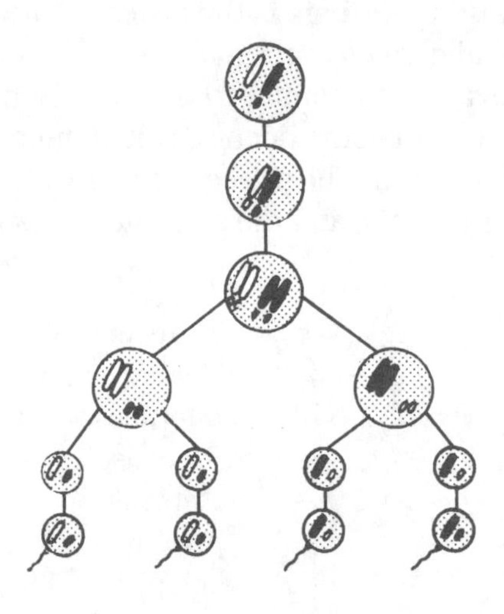

HINT

To remember the difference between mitosis and meiosis, recall that

Mitosis happens in all *body cells*
T for Toe—part of your body—body cells
Meiosis happens in all *sex cells*
S—sex cells

Comparisons	Mitosis	Meiosis
Type of cells	Body cells	Sex cells
Series of changes	One	Two
Results in	Two identical daughter cells	Four sex cells with half the chromosomes of parent cells

BRAIN TICKLERS
Set # 17

1. How many sex cells are formed during meiosis? __________
2. Sperm and egg join together to form a __________.
3. The process by which a sperm and egg join together is __________.
   a. meiosis
   b. mitosis
   c. fertilization
   d. diffusion
4. How many parents are involved in asexual reproduction?
5. Genetic material is found in the __________ of a cell.
6. Genetic variation occurs in __________.
   a. mitosis
   b. sexual reproduction
   c. asexual reproduction
   d. osmosis
7. Replication is another word for making __________.
   a. a copy of DNA
   b. new DNA
   c. less DNA
   d. pieces of DNA
8. List three differences between mitosis and meiosis.

(Answers are on page 88.)

WRAPPING UP

- Cells are the smallest individual beings that have specific tasks or jobs. Some cells can exist on their own; most work with other cells to form tissues, organs, and organisms.
- The parts of the cell are called organelles. Organelles have specific functions that help the cell exist.
- Organelles studied in this chapter were cell membrane, endoplasmic reticulum, nucleus, ribosome, mitochondria, golgi bodies, lysosome, centrioles, nucleolus, chloroplast, cell wall, vacuole, and cytoplasm.
- The plant cell has a cell wall and chloroplasts. The animal cell does not.
- The cell theory is the foundation of all cell biology.
- Scientists use microscopes to study materials that are too small to see on their own. The field of vision is the area that is viewed through the microscope's ocular lens.
- Diffusion is the movement of particles from an area of high concentration to an area of low concentration.
- Osmosis is how particles move across water from an area of high concentration to an area of low concentration.
- Equilibrium occurs when the particles are at an even state and no longer need to move due to difference in concentration.
- Active and passive transport are the terms given to the energy requirements for materials to move. Active transport needs energy to move; passive transport does not.
- Cellular energy is produced through a process called cellular respiration. The cell takes in raw materials—glucose (sugar), water, and oxygen—to create ATP and give off carbon dioxide as a waste product of animals.
- Plants use the carbon dioxide given off by animals along with the raw materials of water and energy from the sun to go through the process called photosynthesis.
- Photosynthesis takes place in the chloroplasts of the plant. The end products are glucose, oxygen, and water. This oxygen goes into the environment for animals to breathe.

- Cells reproduce to form identical daughter cells through a process called mitosis. This is considered asexual reproduction.
- Sex cells go through an additional series of changes called meiosis, which results in four cells with half the genetic information as the parent.

BRAIN TICKLERS—THE ANSWERS

Set # 11, page 52

1. Hooke, cut cork, cells
2. animals
3. c. all cells are surrounded by a cell wall

Set # 12, page 57

1. To increase the magnification of an object while using the compound light microscope, change the objective lens to a higher power.
2. electron microscope
3. ocular lense or eyepiece
4. stereo
5. c. high power
6. low power

Set # 13, page 62

1. nucleus
2. endoplasmic reticulum
3. organelles
4. chloroplasts and cell walls
5. mitochondria
6. A, cell wall; B, cell membrane; C, vacuole; D, nucleus; E, chloroplast

Challenge: The cell can be compared to a school by assigning the different organelles to the parts of the school community: nucleus, principal; ribosomes, teachers; endoplasmic reticulum, hallways; golgi bodies, custodians; mitochondria, furnace/ boiler room or cell transport

Set # 14, page 67

1. c. cell membrane
2. a. lipids and proteins
3. c. is selective in what passes through it
4. a. energy is needed for the materials to move across the membrane
5. b. of osmosis
6. c. equilibrium
7. a. diffusion
8. b. false feet
9. a. how materials move into and out of the cell
10. a. move with ease

Set # 15, page 72

1. d. chloroplast
2. b. cell wall
3. a. carbon dioxide and water
4. b. ATP
5. c. cytoplasm
6. a. mitochondria
7. c. glycolysis
8. d. roots
9. lactic acid
10. Animal cellular respiration takes place in the cytoplasm and the mitochondria. Oxygen and glucose are needed for ATP to be produced. Plant cellular respiration is called photosynthesis. Sunlight is absorbed by the leaves and is used to create food. Water is absorbed by the roots of the

plant for photosynthesis. Water and carbon dioxide are the raw material for photosynthesis.

Challenge:

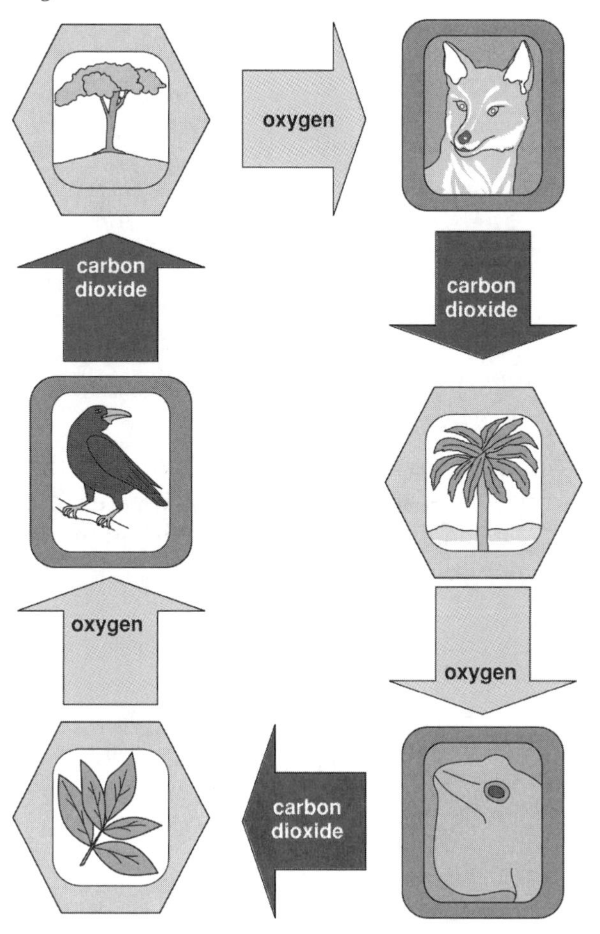

Set # 16, page 78

1. c. growing
2. interphase, mitosis, and cytokinesis
3. d. two identical
4. cytokinesis
5. a. spindle fibers
6. Interphase, prophase, metaphase, anaphase, and telophase.
7. c. metaphase

8. During mitosis, the genetic material replicates.
9. prophase
10. a. anaphase

Set # 17, page 83

1. four
2. Sperm and egg join together to form a zygote.
3. c. fertilization
4. one
5. nucleus
6. b. sexual reproduction
7. a. a copy of DNA
8. Mitosis creates two daughter cells, identical to the parent; there is one series of changes. Meiosis creates four sex cells, with half the genetic material from a parent; there are two series of changes.

CHAPTER FOUR

Plants

PLANT STRUCTURE AND FUNCTION

Plants are a vital part of all life on the earth. Plants aren't always given the credit they deserve. Without plants, animals would not be able to survive. Plants are living organisms that require *air*, *water*, *nutrients*, and *light* to live and thrive. Plants have an amazing design that allows them to grow and survive in different environments. They grow, breathe, change, make food, and reproduce during their lifetimes.

Plants help animals survive by giving off oxygen and making nutrients such as sugar, fruit, and vegetables. They provide animals with many of the substances that they need to survive.

Photosynthesis is a process by which plants make their own *food* and give off *oxygen and water* as waste products. The food made by the plant is a form of sugar called glucose.

Plants contain microscopic little green organelles called chloroplasts, which are made up of chlorophyll. Chloroplasts are located in the cells of the leaf. They are responsible for capturing the sunlight so that the plant can make food.

Oxygen is a gas that is needed by many animals in order to survive. Humans inhale (breathe in) oxygen and exhale (breathe out) *carbon dioxide.*

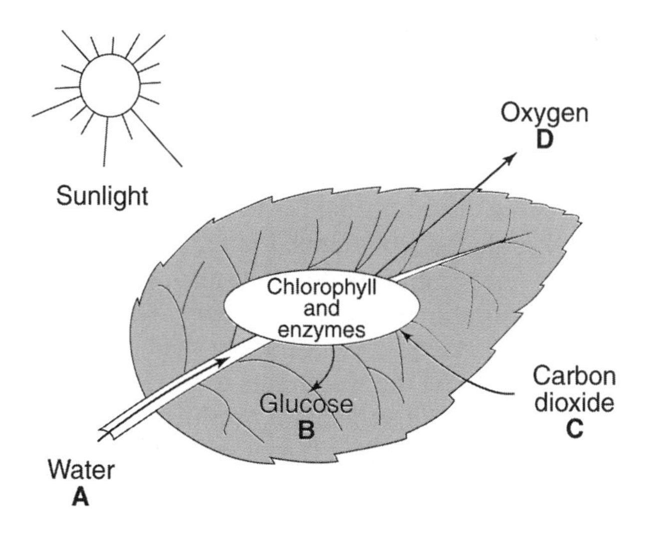

Photosynthesis is defined by this word equation:

$$\text{water} + \text{carbon dioxide} \xrightarrow{\text{energy from sunlight}} \text{glucose} + \text{oxygen}$$

$$6\,H_2O + 6\,CO_2 \xrightarrow{\text{energy from sunlight}} C_6H_{12}O_6 + 6\,O_2$$

The leaves of the plants take in carbon dioxide from the environment. The air moves into the leaves through little openings on the underside of the leaf called **stomates**. They give off oxygen, which animals need to survive. Stomates are found scattered all around the bottom layer of the leaf. These cells open and close to take in and let out gases.

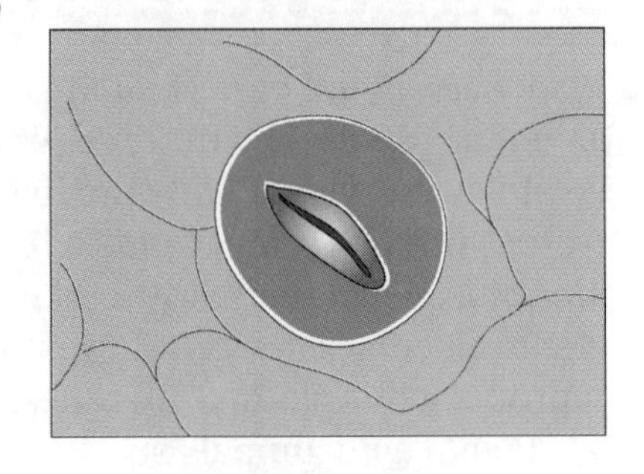

HINT

Stomates work like your mouth, taking in and letting out gases.

Water is vital for all living things. The plant takes in water from the environment through the roots in the ground. Plants absorb water from the soil through their roots. This water can originate from deep in the soil. The water travels from the roots through small tubes in the plant that travel up the stem to the leaves. These small tubes are also called **capillaries**. Water is pulled up the plant due to a process called **transpiration**. Transpiration is the evaporation of water into the atmosphere from the leaves and stems of plants. Plants pump the water up from the soil to deliver nutrients to their leaves. This pumping is driven by the evaporation of water through the stomates. Transpiration accounts for approximately 10 percent of all evaporating water. When the water gets to the leaves, it is used in photosynthesis and helps the plant produce food.

Light is provided by the sun. It is absorbed by the leaves and is necessary for the plant to produce food. Sunlight provides the energy for the entire process of photosynthesis.

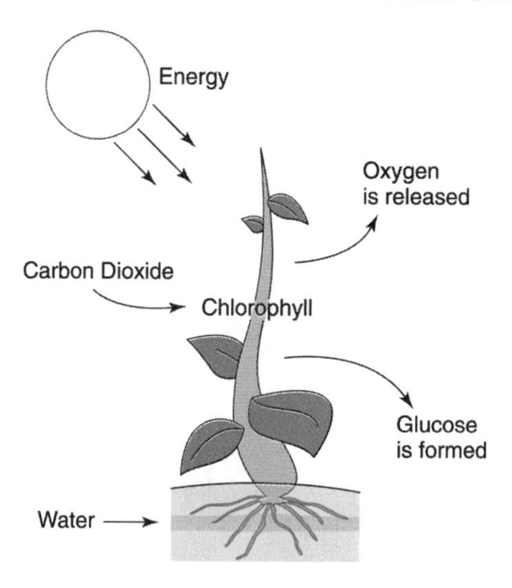

Photosynthesis

BRAIN TICKLERS
Set # 18

1. In order for plants to live and thrive, what things do they require?
   a. food and water
   b. air and light
   c. air, water, and light
   d. light only
2. Chlorophyll is used for __________.
3. What kind of plants and or plant products do you like to eat? Give at least three examples.
4. What are some ways that an animal might use plants to survive? Give at least two examples.

5. What is the process that plants do to make food?
    a. photosynthesis
    b. respiration
    c. digestion
    d. propagation
6. When plants make food, they give off __________ and __________.
7. Where does the plant get the energy to go through photosynthesis?
    a. roots
    b. sunlight
    c. carbon dioxide
    d. leaves
8. What part of the plant absorbs carbon dioxide?
9. What part of the plant takes in (absorbs) water?
10. What part of the plant absorbs sunlight?
11. Describe what transpiration does to help the plant survive.

(Answers are on page 104.)

PARTS OF THE PLANT

There are four major parts of a plant: root, stem, leaves, and flower. Each part of a plant has a specific job (function).

Roots are the part of the plant that branches into the soil. They are responsible for absorbing water and nutrients from the soil. Many roots have thin hair-like growths called root hairs that extend from the main root. They are very thin tissues usually only one or two cells thick. These root hairs increase the amount of water a plant can absorb.

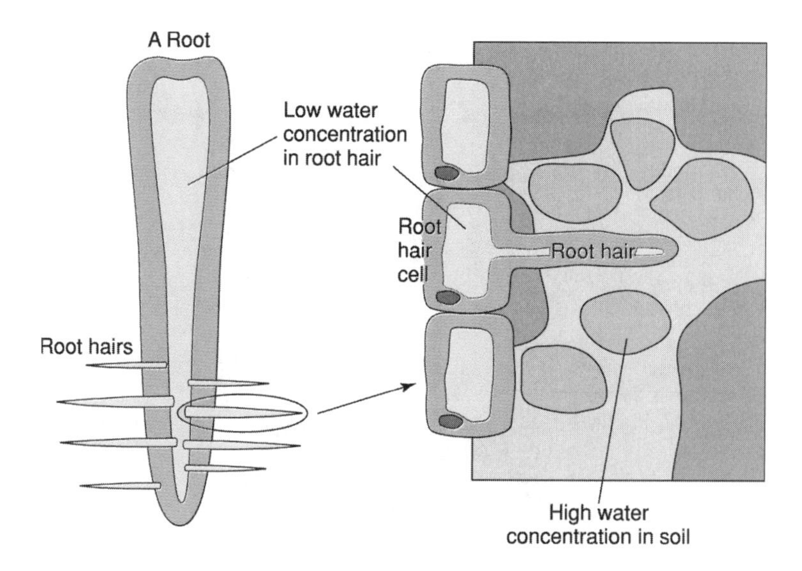

The roots also help anchor the plant in the soil; the larger the plant the deeper and stronger the root.

Stems, stalks, and trunks are the part of the plant that extends from the roots up to leaves and flower. The stem provides the pathway of water and nutrients from roots to leaves and flower. The stem supports and holds up the plant. It is also the pathway for transporting to the roots the extra food (glucose) that was made in the leaves but that the plant does not immediately need so that it can be stored for later.

The food-making factory of the plant is the leaf. **Leaves** extend from the stem; they capture sunlight, regulate the passage of gases through the stomate, and are the site of photosynthesis.

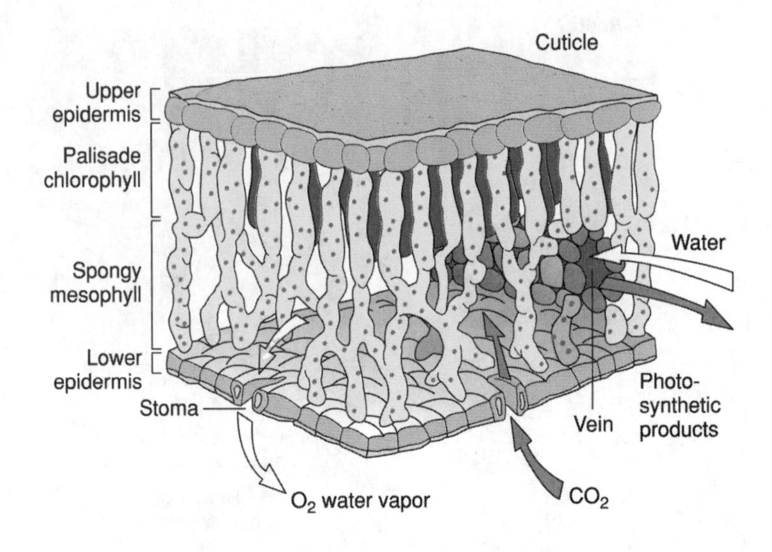

If you took a leaf, cut it in half, and looked at the layers of the leaf from the top to the bottom, it would look like this diagram. The leaf is made up of layers that have specific jobs:

Cuticle—a thin waxy layer responsible for protecting the leaf from getting burnt by the sun.

Upper epidermis—tightly organized cells give shape to the leaf and capture the sun's energy. It can then be used in photosynthesis.

Palisades layer—large amounts of chlorophyll in this layer work hard performing photosynthesis.

There are specific parts of the leaf that help in photosynthesis:

Spongy layer—Just as its name suggests, the cells in this layer are loosely arranged so that air and water can be exchanged.

Vein—The lines that you see along the leaf bring water to the leaf and glucose back down to the rest of the plant.

Lower epidermis—These tightly organized cells with stomates are located throughout the layer for gas exchange.

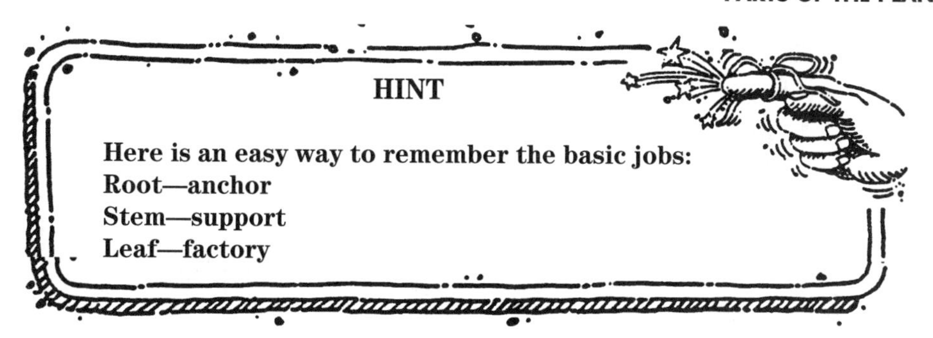

HINT

Here is an easy way to remember the basic jobs:
Root—anchor
Stem—support
Leaf—factory

The **flower** is usually the favorite part of a plant and what we remember them for. The flower is found along and on top of the stem. It is responsible for reproduction. With the help of animals (insects and birds), rain, and wind, flowers can be **pollinated** so that new seeds can be formed that will grow into new plants.

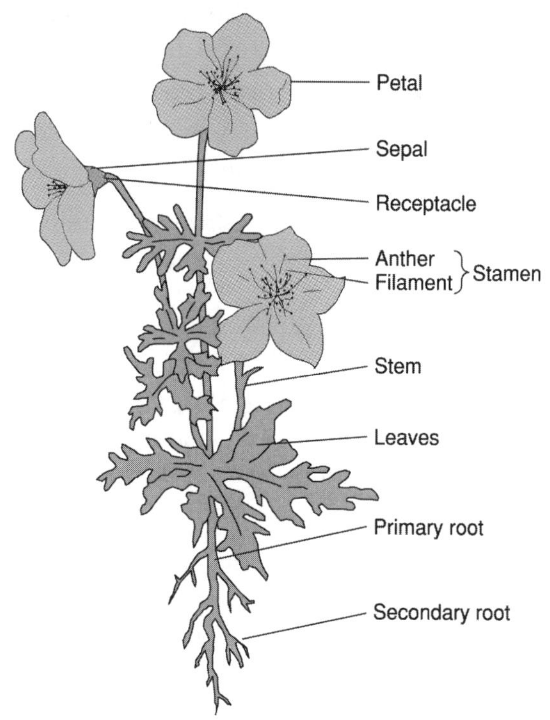

BRAIN TICKLERS
Set # 19

1. The two functions of the roots are __________.
   a. support and reproduction
   b. food-making factory and anchor
   c. reproduction and transport
   d. anchor and absorb
2. Describe the job (role) of the stem in photosynthesis.
3. What part of the plant do we call the food-making factory? Why?
4. How does the flower help the plant reproduce?
5. The stomates of the plant have a unique job. Describe the structure and function of the stomates.

(Answers are on page 104.)

PLANT ADAPTATIONS

Some living things are able to change with the changes in their environment. These changes are called **adaptations**. Adaptation is an organism's ability to adjust to its surroundings. Some plant adaptations might include

- Thorns on stems to protect the plant from being eaten
- Flowers with pleasant scents to attract birds and insects for fertilization
- Large leaves to increase surface area for photosynthesis
- Colored or patterned leaves and/or flowers to blend into the environment
- Roots that grow longer to reach the water supply (For example, corn plants have roots that are 2.5 meters deep, while some desert plants have roots that extend 20 meters into the ground.)

- Leaves and stems that develop tough outer coatings to protect the plant from the environment
- Stems that become thicker to prevent water loss

Over time plants have been able to adapt to their environments in special ways. Think of different climates and the types of trees or plants that live in those regions.

Tropical regions are known for being very hot; they have rain but due to the heat, the water evaporates very fast. In the Tropics, some of the trees and plants have large leaves; they create a covering or canopy over the bottom of the tree so that the area below is cool and the roots can absorb water without it being evaporated too quickly in the warm weather.

In the desert, the cactus is able to survive, even though there is little rain. A cactus has a very involved system where it processes and stores water deep inside the cactus until it is needed, similar to a camel.

In those areas where there are extreme differences in the temperature, plants go through cycles with the seasons. Trees in the middle of the United States (from coast to coast) bloom in spring, are full of leaves for the entire summer, have leaves that change colors in the fall, and then fall off so that in the winter the tree is **dormant**—sleeping so that it doesn't get damaged by the cold weather, ice, and snow. Trees that go through this type of cycle are known as **deciduous**.

In Northern and Arctic regions, the trees have adapted to survive in extreme colds by being what is known as evergreens or **conifers**. These types of trees stay green all year long; their chlorophyll is found in needle-shaped leaves.

The root systems of plants adjust to what is best for them in a certain environment. A large tree will have roots that go deep and wide; this provides the tree with a considerable amount of support. Other plants that don't need to be so strong have thin root systems that bring up the water. Lichens and moss are a type of plant that grows on rocks and the outside of trees. Their roots are microscopic and can bury into the rock to keep the plant on it.

Some plants are able to "hibernate" for the winter, after they have gone through their flowering season; their leaves fall off and the plant essentially dies, but they re-grow the next spring. These plants are called **perennial**. **Annual** plants only live for one season and have to be replanted every spring.

BRAIN TICKLERS
Set # 20

1. Give one adaptation a plant can have and explain how it helps the plant.

(Answers are on page 105.)

PLANT REPRODUCTION

The **flower** is the reproductive part of the plant. Flowers are usually brightly colored and have a nice scent for a reason. These colors and smells attract birds and insects and help with the reproduction process. Flowers go through sexual reproduction; it is called this because there are male and female cells being joined to form the offspring (seed). Pollination is the process by which a plant combines its male and female cells to create offspring in the form of seeds.

In the flower, there are male and female reproductive parts.

- The male organ of the flower is called the **stamen** (containing pollen). At the top of the stamen sits many cells of pollen. The pollen has to be moved to the female for fertilization to occur.
- The female organ of the flower is called the **pistil** (containing eggs). The top of the pistil is a sticky platform called the **stigma**. The pollen lands on the stigma and then travels down the **style** to the **ovary**.

Seeds are formed when cells from the male and female parts of the plant combine. Seeds grow in the center of a flower and continue to develop there after the petals fall off the plant. If the seed surroundings swell up, they can form a **fruit**. All fruits have seeds. Some have one big seed, called the pit, as with a peach. Some fruits can have a few seeds, like the apple. Some fruits have many little seeds, like a kiwi.

Seeds can be spread out or dispersed by a plant's own mechanism and/or in a variety of ways that can include wind, water, and animals. Seeds contain stored food that aids in germination and the growth of young plants. When plants reproduce sexually, the offspring has characteristics similar to those of the parents.

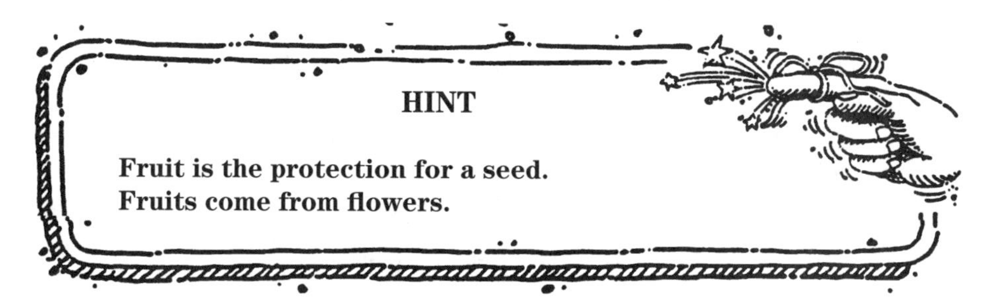

HINT

Fruit is the protection for a seed.
Fruits come from flowers.

The seed is considered to be the embryo of a plant. It stays at rest and doesn't grow until the environment is just right for it to emerge. This process by which the seeds grow into new plants is called **germination**. Seeds will begin to germinate when the soil temperature is in the appropriate range and when water and oxygen are available. The temperature and soil conditions are specific to the seeds. That is why we see different plants growing at different times of the spring and summer.

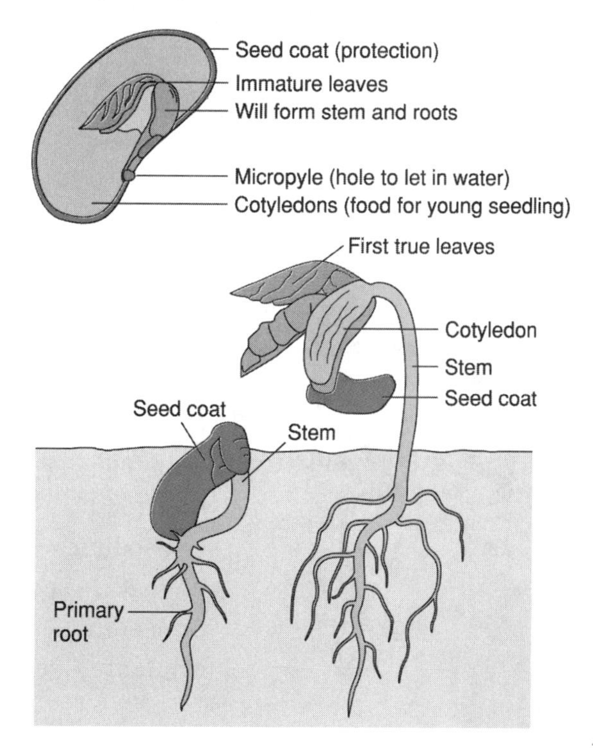

Seeds contain their own food source called the **cotyledon**. This stored food helps them grow and survive until they are able to produce a leaf and begin to make their own food.

LIFE CYCLES OF PLANTS

Plants and animals have life cycles that may include the beginning of life, development into an adult, reproduction as an adult, and eventually death. The length of time from the beginning of the plant to its death is called its life span. Each kind of plant goes through its own stages of growth and development. They may include seed, young plant, and mature plant.

All plants experience growth when they increase in size. The overall size of some plants depends on factors in their environment such as amount of space, light, and water.

Life Cycle of a Flowering Plant

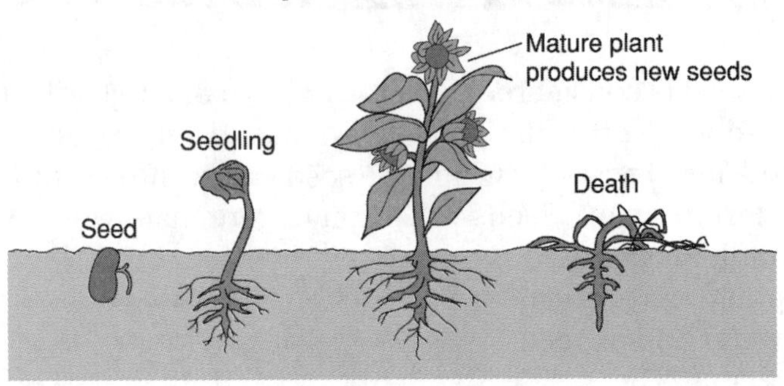

BRAIN TICKLERS
Set # 21

1. How does the fruit develop?
2. Which is an example of a typical life cycle of a plant?
   a. seedling → seed → plant
   b. seed → seedling → plant
   c. seedling → plant → seed
   d. plant → seedling → seed
3. Seeds are usually found inside a fruit. How can a fruit help in distributing seeds?

4. The part of the seed that provides nourishment is the __________.
   a. seed coat
   b. cotyledon
   c. hypocotyls
   d. root
5. Give two factors that have an effect on the germination of a seed.
6. How does a plant sexually reproduce? What is the process called?
7. Give one example of how a plant that lives in the desert might be adapted to survive in the dry environment.

(Answers are on page 105.)

WRAPPING UP

- Plants are living organisms that are made up of cells. The plant requires air, water, nutrients, and light to survive.
- The parts of a plant have specific functions such as the roots, which act like an anchor; the stem, which transports materials; the leaves, the site of photosynthesis; and the flower, the site of reproduction.
- Plants are able to make their own food through the process of photosynthesis.
- Photosynthesis takes place in the chloroplasts of the leaf. The plant takes in carbon dioxide and water. Plants use energy from the sun to create glucose and give off oxygen.
- The air moves into and out of the leaf through openings under the leaf called stomates.
- Transpiration is the process by which water travels up the plant.
- Plants reproduce both asexually and sexually. The flower is the site of sexual reproduction. The new embryo created is called the seed.
- The process by which a seed grows into a plant is called germination.

BRAIN TICKLERS—THE ANSWERS

Set # 18, page 93

1. c. air, water, and light
2. photosynthesis
3. Answers will vary. Possible answers include vegetable, fruits, and wheat.
4. Answers will vary. Plants are a source of nourishment or food for animals. Animals inhale the oxygen that is given off by the plant.
5. a. photosynthesis
6. oxygen and water
7. b. sunlight
8. Gases enter the leaf through the stomates of the leaf.
9. Water is pulled up the plant by the roots.
10. The cells of the leaf absorb sunlight.
11. Transpiration is the process by which water is pulled up the stem of the plant from the root system.

Set # 19, page 98

1. d. anchor and absorb
2. The stem is responsible for carrying water up the plant from the roots to the leaves and nutrients down the plant and supports the plants.
3. Leaves are called the food-making factory because they contain chloroplasts, the site of photosynthesis.
4. The colors and scents bring insects and birds to fertilize the plant.
5. Stomates are the cells on the bottom of a leaf that allow gases into and out of the leaf.

Set # 20, page 100

1. Answers can vary: Roses have thorns for protection, cactus can hold water to survive in the desert, palm trees have wide leaves to absorb extra sunlight, and so on.

Set # 21, page 102

1. Fruit develops around the seed and comes from a flower. Some examples of fruits are apples, oranges, peaches, watermelon, cantaloupe, and grapefruit.
2. b. seed → seedling → plant
3. Seeds can be spread by wind, water, birds, and insects. They can also come from fruit. When the fruit is moved by animals or falls to the ground and is broken open, the seeds can fall to the ground and develop into new plants.
4. b. cotyledon
5. Factors that have an effect on the germination of a seed include temperature, moisture, soil conditions.
6. Plants sexually reproduce when the pollen from the anther travels to the stigma of the flower; it then travels to the ovule, where it joins and creates an embryo, also known as a seed. This process is known as pollination.
7. Answers may vary. Plants that live in the desert might develop deep root systems to get water; cactus store large amounts of water in their thick stems and have few or no leaves.

Animals

Animals are living organisms that need air, water, and food i order to live and thrive. All animals grow, take in nutrients, breathe, reproduce, eliminate waste, and die. Each animal has different structures that perform different functions in their growth, survival, and reproduction.

All animals are **heterotrophs**. Heterotrophs can't make their own food from sunlight, carbon dioxide, and water. Heterotrophs have to eat something to get energy. They have to eat an **autotroph** or another heterotroph. Then they digest what they ate and turn it into nutrients for energy. Animals that eat only plants (most plants are autotrophic) are generally called vegetarians or herbivores. Animals that eat mostly other animals are called carnivores. Humans and other animals that eat both plants and animals are called omnivores.

The jaw and teeth of an animal can tell you what it eats. Carnivores have sharp long fangs for tearing of meat; herbivores have wide teeth for grinding and grazing. Omnivores have a combination to help them eat both plants and animals.

Most animals are covered with hair or fur. The hair traps the air to keep them warm. The hair can also be on the face as whiskers; those hairs are then used for sensory reasons.

Growth is the process by which plants and animals increase in size. All living things are constantly going through some form of growth and repair. When they are young, their overall bodies are growing and developing. This is evident by how tall or big the organism gets. As an organism moves into its adult stage of life, it continues to grow. This is proven by how its hair and nails continue to grow and need to be maintained. When an animal breaks a bone, it is able to heal over time. This healing is done by the body growing new bone cells to repair the break.

You have probably witnessed this on your own body. Think about some time when you cut yourself and after a short period of time your cut healed. Most of the time, your body can heal from a cut without any scar. This happens because your body is always growing/making new skin cells.

Animals have different ways by which they can move around. Birds use their *wings*, fish use their *fins*, and many other animals such as horses, dogs, bears, and humans use their *legs*. Animals need to move around so that they can find a safe place to live/shelter, to find food, and to escape predators.

Lobsters and crabs protect themselves with their *claws*, turtles have *shells*, the porcupine has *spines*, and birds use their *feathers*. Others like bears and beavers have *fur* and other animals use *scales* to help protect them from predators.

Some animals are able to change the color of their body coverings to match that of their surroundings. We call this **camouflage**, and it helps animals protect themselves from predators. Think of a lion in the wild; his golden fur enables him to blend into the desert landscape. Chameleons are lizards that can change their colors based on the color of their surroundings. If it is on the sand, it will turn a tan color; if it is on grass, it will turn green; this allows them to hide from predators.

A **defense mechanism** is a way that an animal protects itself. Some animals protect themselves by sending out special *sounds* or *smells;* this is another type of **defense mechanism**. These smells and sounds can attract other animals so that they can mate, or they can help them fight a predator. Skunks are able to give off a very foul smell when they are in danger. This is how they protect themselves.

ANIMAL RESPONSES TO THE ENVIRONMENT

Behavior is the way that organisms respond to changes in their environment or to a stimulus. Humans shiver when it is cold; by doing so, we create more heat in an attempt to warm our bodies. In the hot weather, our bodies sweat **(perspiration)** bringing water to our skin; when air brushes across our wet skin, it cools down our body helping us cool off.

When we get nervous, scared, or upset, our bodies respond; we increase our heart rate, breathe faster, and begin to sweat. This is how our bodies prepare for whatever danger may be coming.

Seasonal changes

Some animal behaviors are influenced by environmental conditions. These behaviors may include nest building, hibernating, hunting, migrating, and communicating.

Many animals are able to change their bodies with the seasons. We wear winter coats, gloves, and hats in preparation for going into the cold weather. Some animals grow thicker fur in the winter and shed it in the warmer months; this helps their bodies regulate their body heat. Other animals change their amount of body fat with the seasons. Body fat is a form of stored energy and helps them get through long periods of time with little food. Squirrels collect acorns in preparation for the long winter.

Some animals live a very different life in the warm weather than in the cold weather. Some animals go into **hibernation** during the winter months; their bodies slow down and go into a dormant state. Bears go into hibernation, but before doing so, they eat a great deal of food to increase their body fat so they will be able to go for a long period of time with little food and sleep until the cold winter months pass.

Moving to a warmer climate in the cold months is another way some animals deal with changes in the weather. This is called **migration**, another type of seasonal change. Birds are known for flying south in the winter to avoid the cold.

BRAIN TICKLERS
Set # 22

1. Heterotrophs __________ food, and autotrophs __________ food.
2. List and describe two ways that animals respond to changes in their environment, such as seasonal changes.
3. When animals move to a warmer climate to avoid the change in seasons, we call this __________.
   a. hibernation
   b. camouflage
   c. perspiration
   d. migration

4. Some animals prepare for the long winters by storing food and going dormant. This is called __________.
   a. hibernation
   b. migration
   c. camoflage
   d. conservation
5. Explain the differences between how dogs, birds, and fish move around.
6. Some animals have claws, spines, or shells or give off a smell as a __________.
   a. migration
   b. hibernation
   c. seasonal change
   d. defense mechanism
7. Animals that are able to change color or are of a color similar to the environment are said to be able to __________ themselves so that they cannot be easily seen.

(Answers are on page 120.)

ADAPTATIONS

Individual organisms and species change over time. Throughout time, plants and animals have changed depending on their environment. They do this over long periods of time and pass the adaptations from one generation to the next so that their species can survive and flourish. The saying "survival of the fittest" means that those who are best adapted will survive over those who are not.

Camels are able to go for long periods of time with little water; this makes them the best type of animal to live in the desert. Cactus plants are also able to go for long periods of time with little water. That is why they survive in the desert.

Compare the skeletal system of a cat and a horse. They are clearly two different genera and species, but they are extremely similar. In **comparative anatomy**, we review how the body structures of different organisms are similar. Look at the similarities and differences between animals' arms/wings/legs.

Depending on how the organisms live, their arms/wings/legs adapted to help them survive.

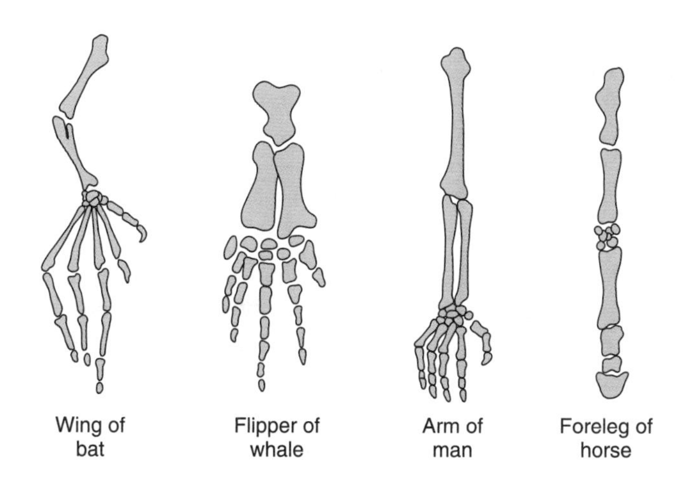

These similar body structures are known as *homologous structures*.

The beak of a bird is similar to human lips and teeth. A bird's beak helps it capture and take in food. Compare the various beaks and think about what type of food each bird eats. The shape, strength, and design of the beak vary depending on whether it is part of a bird of prey, a seed-eater, or a fish-eater.

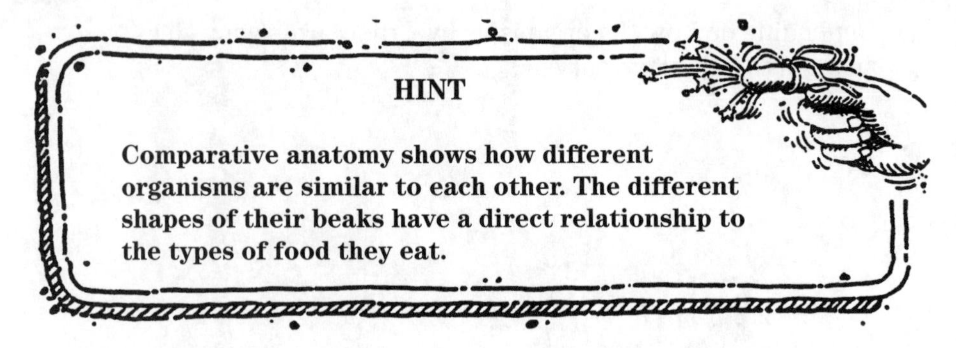

HINT

Comparative anatomy shows how different organisms are similar to each other. The different shapes of their beaks have a direct relationship to the types of food they eat.

Elephants have long trunks that enable them to grab and manipulate food and water. They also have very large ear flaps that help them hear. Because of their size, elephants move very slowly so having very good hearing helps them protect themselves from predators.

Comparative embryology is the study of the similarities and differences that occur in the development of animals or plants of different orders. Most organisms look very similar in the early stages of embryonic growth. Here is a chart showing how young embryos look alike and then change into the different organisms.

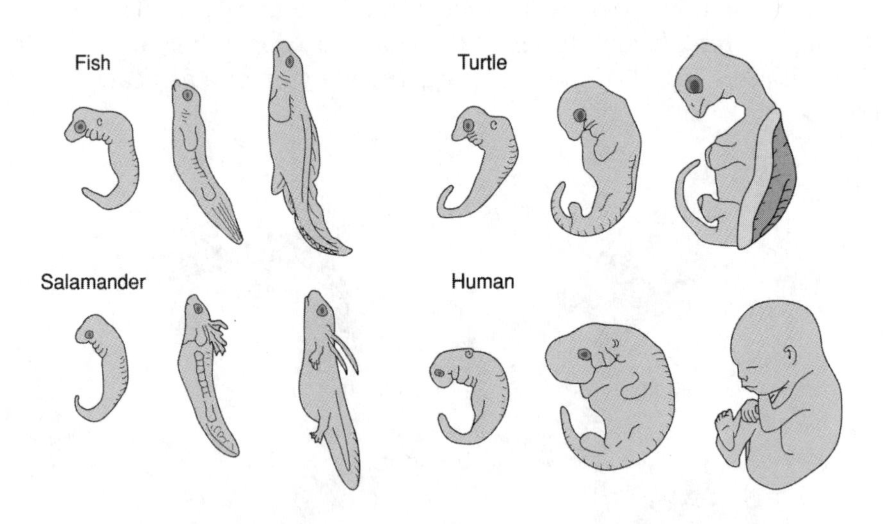

All individuals have **variations**, and because of these variations, individuals of a species may have an advantage in surviving and reproducing. There are many different types of animals within the same species; these differences are called **diversity**. There are many different breeds of dogs, but even within the same breed there are variations. You could have three beagles, and all three could look similar but their markings would be different.

BRAIN TICKLERS
Set # 23

1. A giraffe's long neck helps it eat leaves that are high on trees. This trait can be considered a(n) __________.
   a. adaptation
   b. convenience
   c. camouflage
   d. migration
2. Explain why the pelican's beak is helpful to the bird.

3. What type of environment do animals with gills live in?
   a. desert
   b. forest
   c. prairie
   d. water
4. Explain how animals' limbs adapt to their behavior?

5. Cats have distinct colors and patterns that make them look different from other cats. This is an example of __________.
   a. adaptation
   b. camouflage
   c. variation
   d. hibernation
6. Proof of how animals have adapted for survival in their environment is demonstrated in this drawing of a whale flipper and a human arm. Explain what this statement means.

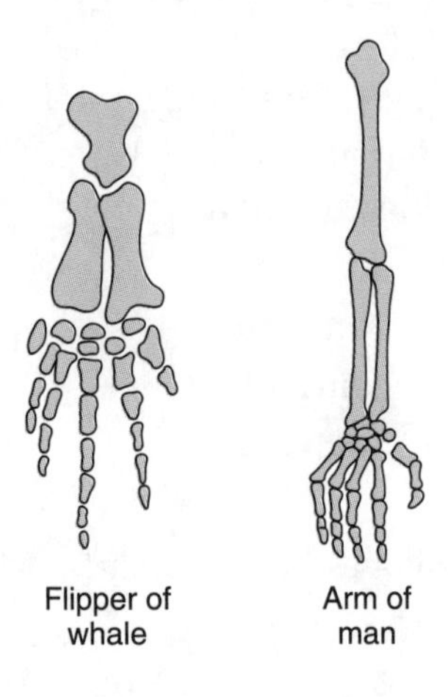

(Answers are on page 120.)

REPRODUCTION/LIFE CYCLES

All living things go through different stages in life. They are born, grow (develop), have babies (offspring), and die. This is called the continuity of life or **life cycle**. When an organism has an offspring/baby it is called **reproduction**. As they grow and change over time it is called **development**.

Living things must reproduce and have babies so that their species will continue. If they were not able to reproduce, their species would become **extinct**, that is, the species would no longer exist, like the dinosaurs.

All living things have specific patterns of reproduction, growth, and development called stages in its life cycle.

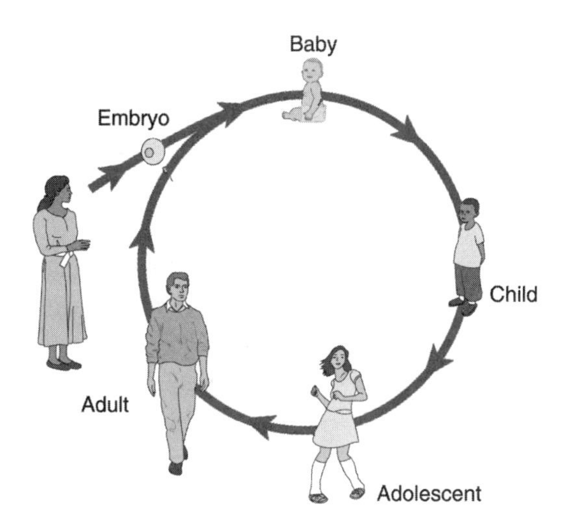

embryo → baby → child → adolescent → adult

Life cycle stages are sequential and occur throughout the life span of the organism. The characteristics of the cycle of life vary from organism to organism.

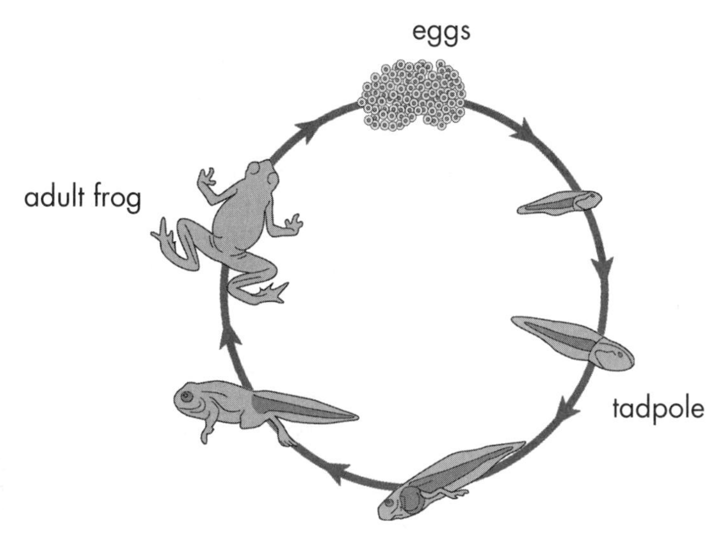

Frog Life Cycle

Remember: Some animals are born as one type of organism, like the frog, beetle, and ladybug, and as they grow, they go into a changing stage called **metamorphosis**.

The length of time from an animal's birth to its death is called its **life span**. The type of environment an animal lives in and the environmental factors that exist can have a direct effect on the animal. Life spans of different animals vary.

Animal	**Life span, average**
Human	79 years
Worker bee	1 year
Rabbit	9 years
Frog	10 years
Turtle	100 years
Elephant	70 years
Parrot	80 years
Ant – worker	6 months
Fruit flies	72 hours

The health, growth, and development of organisms are affected by environmental conditions such as the availability of food, air, water, space, shelter, heat, and sunlight.

Reproduction passes information from parent to offspring. **Asexual reproduction** requires one parent and produces nearly identical offspring. **Sexual reproduction** requires two parents and provides variety in a species. This variety may allow the species to adapt to changes in the environment and help the species survive. A species may change due to the passing of traits naturally or by techniques used and developed by science. Genetic information is passed on in a predictable manner.

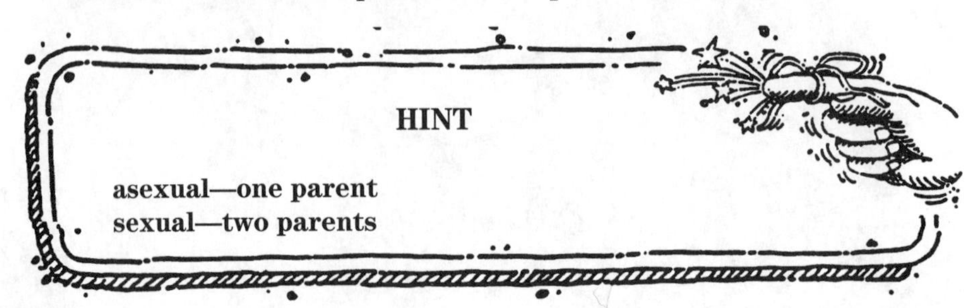

BRAIN TICKLERS
Set # 24

1. Describe the major stages in the life cycles of humans.
2. During which phase of the frog's life span is it able to swim with the help of a tail?
   a. egg
   b. tadpole
   c. larva
   d. adult frog
3. The length of time from an animal's birth until its death is called __________.
   a. life span
   b. instinct
   c. life cycle
   d. development
4. Explain the difference between the life cycle of the ladybug and that of the frog.
5. __________ reproduction results in offspring that are genetically different from their parents.

(Answers are on page 121.)

WRAPPING UP

- Animals are multicultural living organisms that require air, water, and nutrients (food) to survive.
- Animals are considered heterotrophs; they are not able to make their own food.

- All animals grow, take in nutrients, breathe, reproduce, eliminate waste, and die. They have different structures that perform specific functions in order for them to survive.
- Animal behaviors are influenced by environmental conditions; these behaviors include hibernation, migration, nest building, and communicating. They protect themselves with claws, shells, scales, scent, fur, camouflage, and feathers.
- Animal body structure is adapted to help it survive. For example, birds bones are hollow to help them fly.
- Animals' body structures are similar; they are called homologous structures.

BRAIN TICKLERS—THE ANSWERS

Set # 22, page 111

1. cannot make their own, can make
2. Answers may vary. Seasonal changes include hibernation (going into a dormant state for the winter), migration (moving to warmer climates), thick fur (grown in winter to keep animal warm), body fat (increases in winter to keep animal warm), and shivering (shaking to keep the body warm).
3. d. migration
4. a. hibernation
5. Students should discuss how dogs have legs, birds have wings, and fish have fins to help them move.
6. d. defense mechanism
7. camouflage

Set # 23, page 115

1. a. adaptation
2. Pelicans swoop into the water and are able to collect many fish in their large beaks.

3. d. water
4. Students should discuss how the fish has a flipper or fin that is wide like a paddle to help them swim. Birds have wings that are long and angled to help when they fly. Horses have strong legs, and they have hooves that enable them to pound the ground while they run.
5. c. variation
6. The whale's flippers have adapted to create a shorter, wider area so that it can act as a paddle helping the whale to swim. The human needs to use its fingers for many tasks and has defined fingers and long arms to help it survive.

Set # 24, page 119

1. Baby → toddler → child → teenager → adult → elderly
2. b. tadpole
3. a. life span
4. Students can discuss how they both start as eggs and go through metamorphosis for similarities. They can explain how the tadpole swims and the frog then matures to a land animal. The ladybug grows from a larva that crawls to the ladybug who can fly.
5. sexual

CHAPTER SIX

Human Body Systems

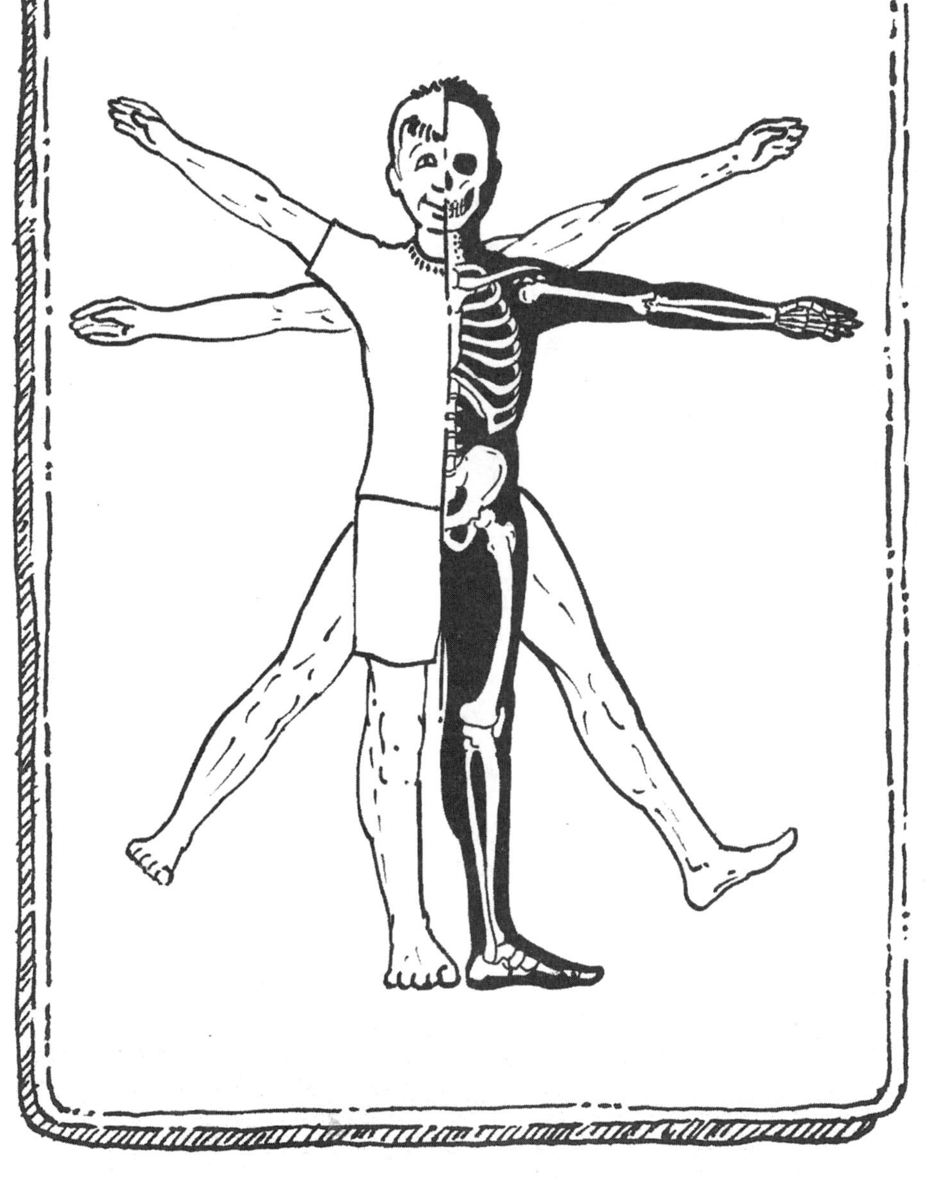

Think of a machine and how all of the pieces have to fit together and work together for the machine to get the job done. The human body is very much like a complex machine; it is made up of small units that join together to form organs. The organs of the human make up organ systems that have specific jobs to do, and in the end all of the organ systems work together for the benefit of the human being.

ORGANIZATION OF ORGANISMS

There is a specific organization to how all animals' bodies are formed. From the smallest unit, all animals are made up of **cells** that have specific jobs. Their shapes and organelles help them perform whatever task they are designed to do. Groups of cells come together to make up **tissues**. Your muscles and skin are examples of tissues. Groups of tissues working together are called **organs**; the lungs are an organ in our body. Different organs working together are called **organ systems**; the lungs work with the diaphragm, trachea, and nasal cavity to form the respiratory system. These organ systems work together to make the **organism**.

cells → tissues → organs → organ systems → organisms

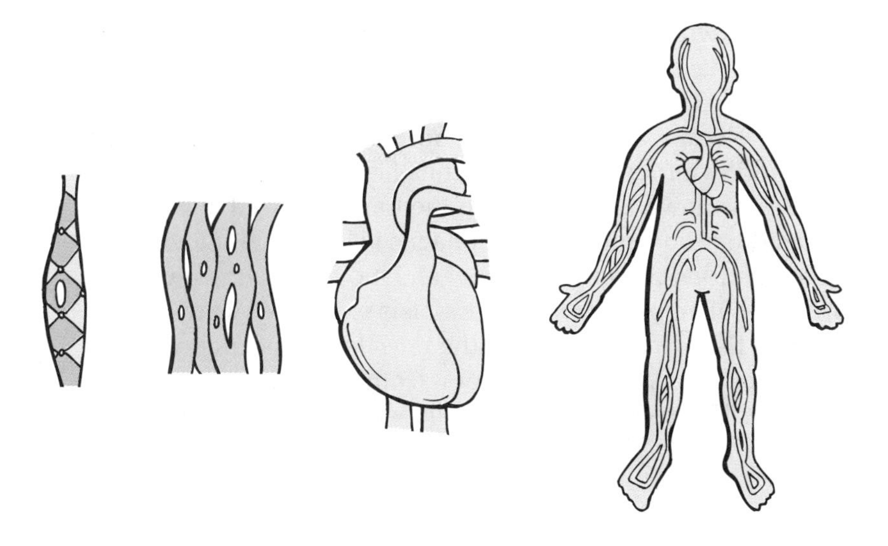

Each organ system has a specific function that it is specially designed to perform. The organ systems do this for the organisms to grow, create energy, and survive.

Let's explore each of the human body systems.

DIGESTIVE SYSTEM

Humans need to eat to get energy for their bodies to perform all of the life functions they need for survival. The **digestive system** deals with the body taking food into it, taking the nutrients from the food, and eliminating all the wastes. These actions are broken down into three phases of digestion.

Ingestion—Food is taken into the body.
Digestion—Nutrients are taken into the body.
Egestion—Undigested food leaves the body.

Humans take in food, water, and nutrients (vitamins and minerals) through their mouths, where their teeth and tongues break up the food into smaller pieces so that it can be swallowed. The saliva glands in the mouth secrete juices onto the food. The saliva begins to break down carbohydrates like bread in the mouth.

Have you ever put a cracker or piece of bread in your mouth and did not chew it right away? Try it! After a few moments of the saliva acting on the bread or cracker, it will begin to taste sweet because the complex sugars found in carbohydrates are beginning to break down.

The chewed food is called a **bolus**. The tongue pushes the broken-down food—bolus—to the back of the mouth so the food can be swallowed. This is when the food is moved into a tube called the esophagus. The esophagus is lined with mucus to make the food easier to swallow and move. Swallowing starts a process called **peristalsis**. Peristalsis is how food moves through the digestive system; the muscles of the digestive system contract to move the food from the esophagus to the stomach. The movement is like squeezing toothpaste out of a tube.

After moving through the esophagus, the bolus enters the stomach where the food is broken down more. In the stomach, **gastric juices** are secreted from the stomach walls. The stomach is a very acidic environment where enzymes break down proteins. The partially digested food then moves into the small intestine where the majority of digestion takes place and the nutrients are **absorbed** into the bloodstream so that they can be transported throughout the body. The small intestine is very, very long so that all of the useful nutrients can be absorbed as the material passes through it. In the very first part of the small intestine, a final secretion comes into play. When the material gets to the end of the small intestine, it is considered waste because the body can't use it. It enters the large intestine, also known as the colon. In the large intestine, the undigested food is now called **feces**. The feces wait in the rectum until the body is ready to release this waste through the anus. At the anus, there is a circular muscle called a sphincter that controls when the body releases the waste. The glands of the digestive system are known as exocrine glands because they deliver the enzymes onto the materials they will act on.

HINT

Folded structures increase the surface area for absorption. The small intestine is very long and folded so that there is plenty of space for nutrients to be absorbed into. When food is chewed, the food is broken down so that there is more surface area for the enzymes to do their work on, thus speeding up the reaction.

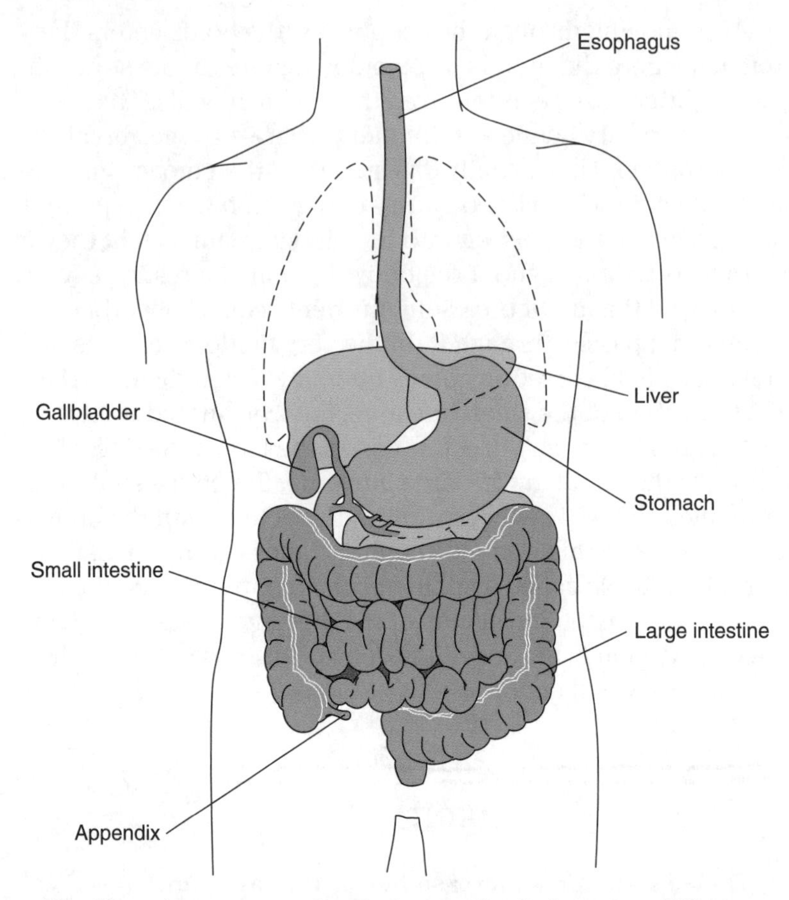

The **liver** is an organ that helps the body in many different systems. The liver is considered a recycling center of the body; it filters the blood, breaks down old blood, and detoxifies the body of chemicals from drugs and alcohol. The liver stores sugar for later use and also produces **bile**, which helps in the digestion of fats. The **gallbladder** is the little sac under the liver that stores bile and then secretes bile into the first part of the small intestine to break down fat.

Digestive enzymes

The chemicals in various digestive system secretions are called enzymes. **Enzymes** are proteins that speed up chemical reactions in the body. Each enzyme has a specific shape that only fits the kind of substance that it breaks down. This is called the **lock and key model**—just like a key, it only works on what it is supposed to work on.

Saliva is secreted in the mouth; it is an enzyme that digests/breaks down starch. Protease/pepsin is secreted in the stomach; it is an enzyme that breaks down proteins.

The substance being digested (broken up) is called the **substrate**. The area where the enzyme and substrate meet is called the **active site**.

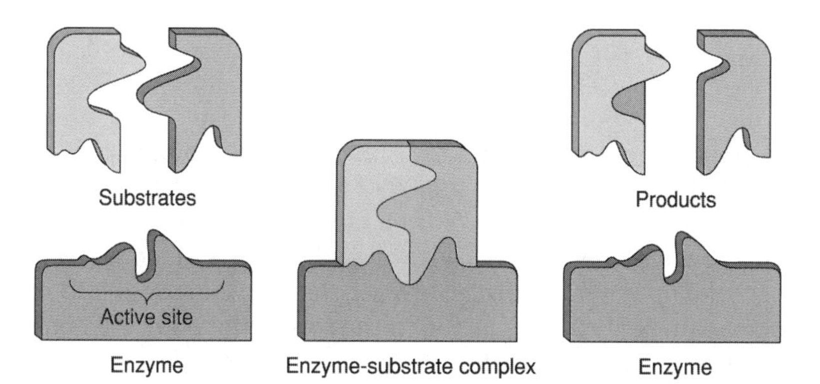

HINT

There are two types of digestion:

chemical digestion—The chemical makeup of a substance is changed. All enzymatic actions are chemical digestion.

mechanical digestion—The shape of a substance is changed. Chewing food by the teeth is mechanical digestion.

RESPIRATORY SYSTEM

Try this: Take a deep breath in through your nose, hold it for a moment then let it out through your mouth. You have just consciously (knowingly) experienced how your **respiratory system** works. Throughout our lifetimes, our respiratory system is hard at work because we need to take in oxygen in order to survive. We do this by exchanging good air, oxygen, for used air, carbon dioxide; this is done by **breathing**. We take in, **inhale**, air that contains oxygen, through the nose into the nasal cavity. In the nose, three things happen to the air. Your nose is a wet environment with little hairs called cilia and lots of blood vessels close to the surface. The hair filters the air and takes out any dirt or dust. The mucus that makes the insides of our noses wet moistens the air before it goes into the body. The blood vessels warm the air. Now that the air has been moistened, filtered, and warmed, it moves from the nasal cavity into the trachea. The trachea is a semi-hard tube that runs from the back of the nose and mouth down to the bronchi. The oxygen-rich air moves through the bronchi and into smaller branches called bronchioles. When the oxygen-rich blood gets to the end of the path, it ends in the alveoli. These alveoli are also called air sacs that make up the lungs. In the alveoli, the oxygen moves into the bloodsteam so it can be carried to the rest of the body. All cells need oxygen to survive. After the cells use the oxygen, they send the waste carbon dioxide back into the bloodstream. Carbon dioxide, which is a waste, leaves the blood with water and passes back into the alveoli; it then travels out of the lungs and is **exhaled** out of the body. This diagram demonstrates how gas exchange occurs in the capillary surrounding the alveoli.

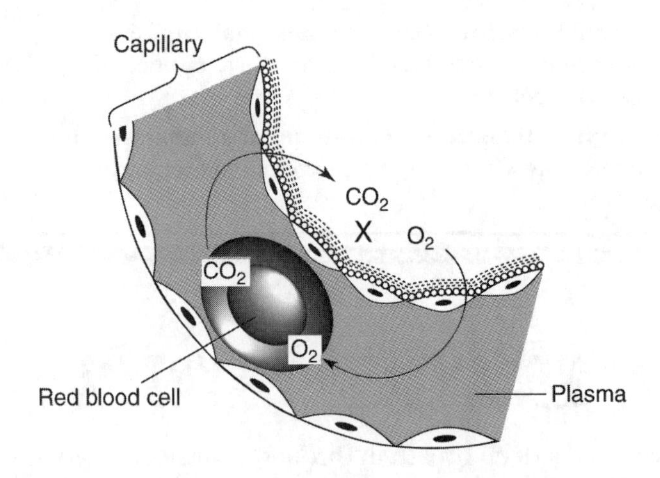

The diaphragm is a large muscle that is found under the lungs. The diaphragm moves up and down as we inhale and exhale and helps the lungs take in and let out air.

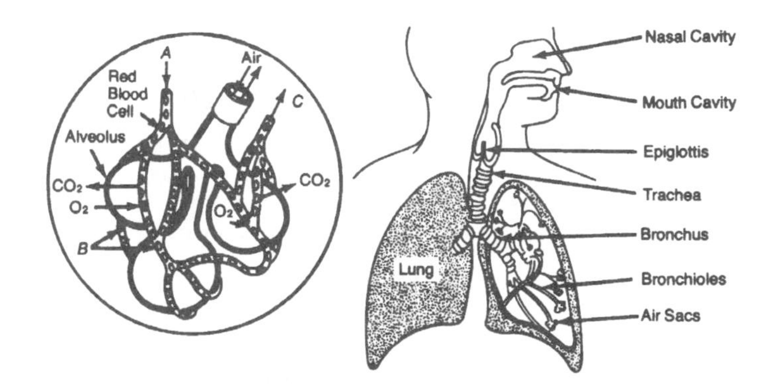

BRAIN TICKLERS
Set # 25

1. What is the function of the digestive system?
2. Mechanical and chemical digestion take place in the __________.
3. The gallbladder secretes __________ into the __________ for the break down of fats.
4. Which organ of the digestive system is the site of absorption of nutrients into the blood?
5. Which organ is not just part of the digestive system?
   a. stomach
   b. liver
   c. small intestine
   d. large intestine
6. Peristalsis begins in the __________.
   a. esophagus
   b. stomach
   c. large intestine
   d. small intestine

7. What is the function of the respiratory system?
8. Describe the path that oxygen takes as it moves from the air outside your body into the alveoli.
9. Explain how smoking could affect respiration.
10. Describe the role of the hair, blood, and moisture in the nasal cavity.

(Answers are on page 156.)

CIRCULATORY SYSTEM

Have you ever tried to feel your heart beat by putting your finger on your wrist to get a pulse? The blood flowing through your body and your heart beating are all part of the **circulatory system**. The body must get nutrients and oxygen to all of its cells, tissues, and organs; it does this by circulating **blood** throughout the body. The circulatory system works as the pump that pushes all the blood around to drop off materials and to pick up wastes and send them to where they will be disposed of. The blood carries the nutrients through tubes called **vessels**. There are three types of vessels:

Arteries—muscular tubes that carry blood away from the heart to the body. The aorta is the largest artery.
Veins—thin tubes that push blood toward the heart. The veins have valves that push the blood forward. The vena cava is the largest vein.
Capillaries—thin tubes, one cell thick, where gas (oxygen and carbon dioxide) and nutrient exchange (glucose).

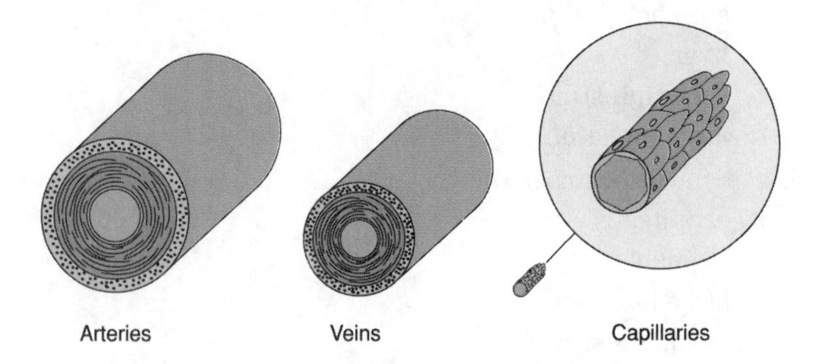

The pathway of blood through the body is as follows:

heart → aorta → artery → capillary → veins → vena cava → heart

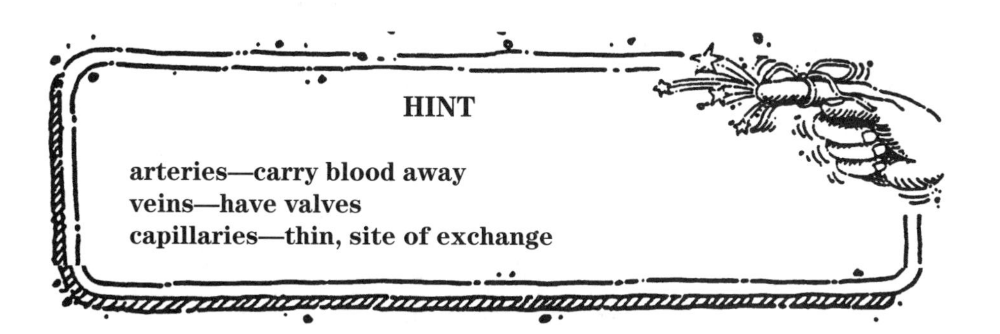

The blood is continuously pumped all over the body because of a strong muscle called the heart.

The **heart** is a muscular hollow organ about the size of your fist that beats around 100,000 times every day. Its main job is to pump the right amount of blood to all parts of the body. The circulating blood brings oxygen and nutrients to the tissues and removes waste.

During a **heartbeat**, the heart first contracts, pushing blood out toward the rest of the body. It then relaxes, filling up with returning blood. How quickly and forcefully the heart pumps depend on the needs of the body. During exercise, the heart must work harder to keep the muscles and organs supplied with all the oxygen-rich blood they need.

Inside the heart are **four chambers**: two on the left side and two on the right. Blood enters the heart and flows into the upper chambers (left and right **atria**). From here it is pushed into the lower chambers (left and right **ventricles**). The ventricles then contract (push together), forcing blood out of the heart.

The left and right sides of the heart have different jobs. The right circulation takes deoxygenated blood that has already traveled around the body and pumps it to the lungs to pick up fresh oxygen. The left circulation takes this newly oxygenated blood and pumps it back out to the rest of the body.

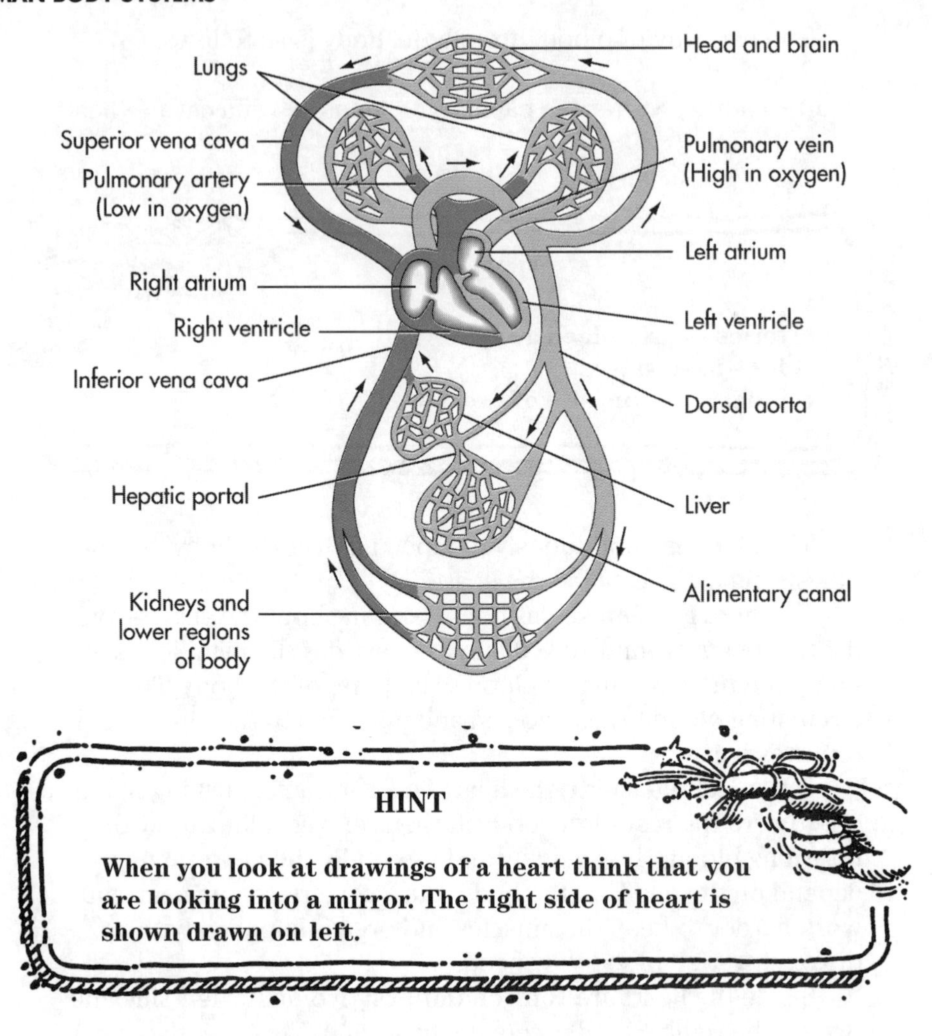

HINT

When you look at drawings of a heart think that you are looking into a mirror. The right side of heart is shown drawn on left.

The different parts of the blood have specific roles.

- **Red blood cells** (RBC) are round cells that do not have a nucleus. They carry oxygen throughout the body.
- **White blood cells** (WBC) are larger cells that have a nucleus; they are the protectors of the body. They fight infections that threaten the body and support the immune system.
- **Platelets** are little particles that travel through the blood and, when necessary, create clots that prevent the body from losing too much blood. If you cut your arm, it bleeds for a minute or two; after a short amount of time you stop

bleeding, thanks to your platelets patching up the opening and creating a clot.

- **Plasma** is the liquid part of the blood that doesn't have cells. It contains antibodies and other proteins.

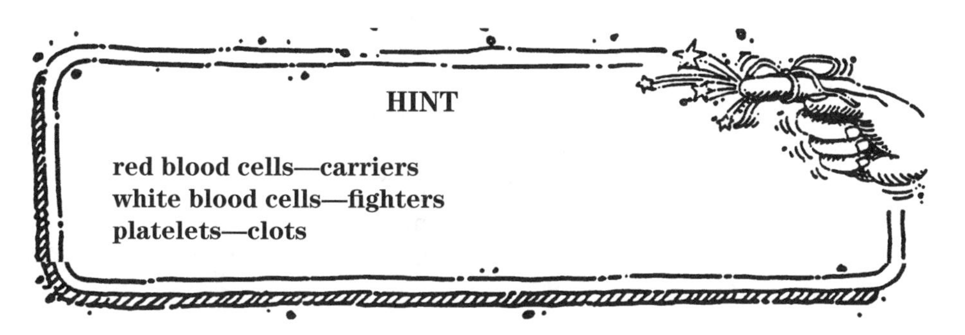

HINT

red blood cells—carriers
white blood cells—fighters
platelets—clots

Blood cells are made in the **bone marrow**. The bone marrow is the soft, spongy material in the center of the bones that produces about 95 percent of the body's blood cells.

Everybody has a **blood type**. The most common blood type classification system is the **ABO** ("A-B-O") system. There are four types of blood in the ABO system: A, B, AB, and O. Your blood type is established before you are born, by specific genes inherited from your parents. You receive one gene from your mother and one from your father; these two combine to establish your blood type. These two genes determine your blood type by causing proteins called agglutinogens, which are found on the surface of all your red blood cells. Agglutinogens are also known as clumping factors.

It is important for people to know their blood type because if they need a transplant there are certain types of blood they can have. O blood type is called the universal donor because it can be given to anyone. AB is the universal recipient; it can take from any type but can only donate to AB.

Immunity is the body's ability to fight foreign substances and cells before they have a chance of making the body ill. Your body has three first lines of defense that try to prevent foreign substances or pathogens from entering the body.

- **Skin**—Destructive chemicals in your skin's oil and sweat may prevent some pathogens from entering the body.
- **Nasal cavity and breathing passage**—Hair, called cilia, and mucus trap pathogens before they enter; they then leave the body by sneezing or coughing.

- **Mouth and stomach**—Saliva and stomach acids can destroy pathogens.

The immune system goes into a "search-out and destroy" mode when it "recognizes" foreign **antigens**. An antigen is any substance that can be bound by an antibody and start some kind of immune response. How does the immune system recognize these foreigners? The body gets ill from an antigen. The body tries to fight it, but it takes time for this to happen, and the person is ill for a short period of time. The body's white blood cells remember what that antigen looks like. If the body gets the antigen again, the white blood cells go to work and fight it before the body gets sick.

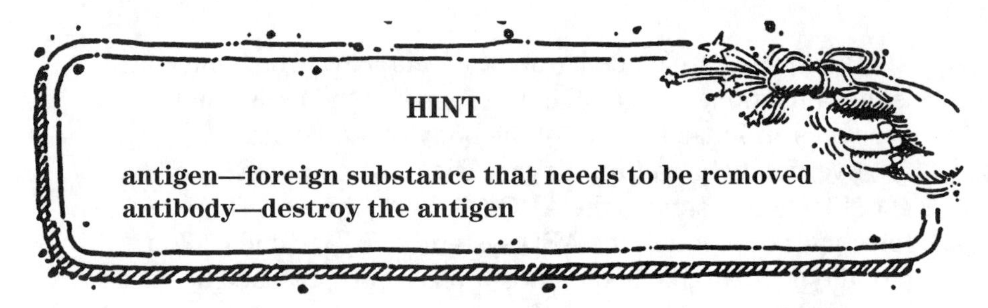

A **vaccine** is a preparation that is used to improve immunity to a particular disease. The person is injected with small noninfectious pieces of a foreign body. The white blood cells identify the foreign body and establish immunity to the foreign body. Smallpox, polio, tuberculosis, diphtheria, and many other diseases have been almost eliminated by vaccinations.

The immune system's first encounter with an antigen is the primary response. A re-encounter with the same antigen causes a more powerful and rapid secondary response. This is called **acquired immunity** or **active immunity** and involves lymphocytes—a type of white blood cell.

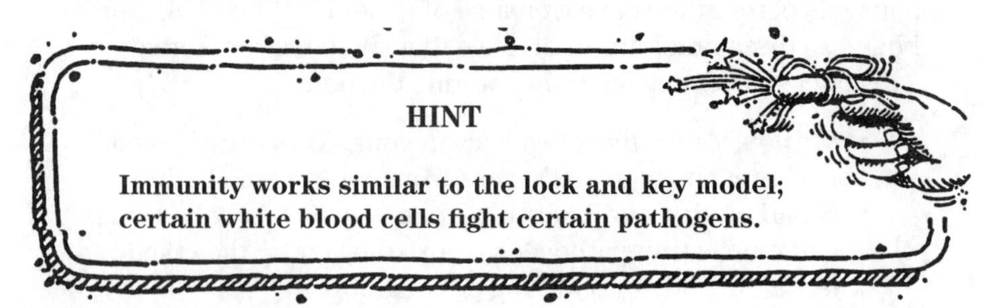

BRAIN TICKLERS
Set # 26

1. Trace the path of a red blood cell from the aorta through the body to the vena cava.
2. What is the difference in the blood traveling through the right and left sides of the heart?
3. What is the function of the circulatory system?
4. Which part of the blood carries nutrients?
   a. white blood cells
   b. red blood cells
   c. platelets
   d. plasma
5. Which is the liquid part of the blood?
   a. white blood cells
   b. red blood cells
   c. platelets
   d. plasma
6. Blood is produced in the __________.
7. There are __________ types of blood; list the types of blood.
8. Describe how a barrier can prevent illness.
9. Why are vaccines helpful in fighting disease?
10. Label the parts of the heart.

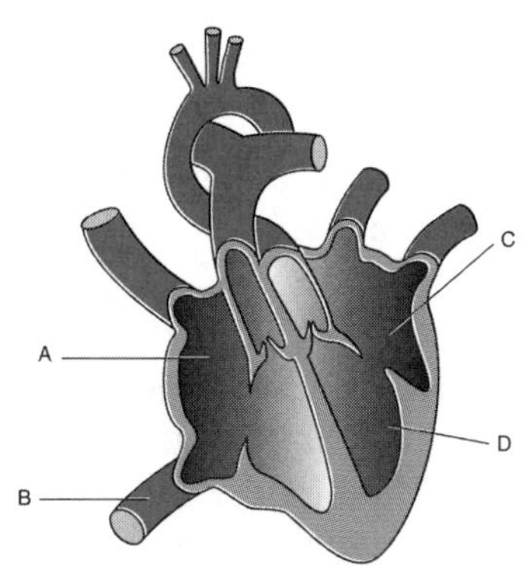

(Answers are on page 156.)

URINARY SYSTEM

What happens to all of the soda, juice, and water we drink? Why do we need to take fluids into our body? As discussed earlier, fluids like water are necessary for all life functions to occur. When the body has used the fluids, the left over liquid waste, as well as the left over chemical waste, leaves the body by the **urinary system**. After the blood carries nutrients and oxygen throughout the body, it needs to get rid of the **chemical wastes**. The blood takes the chemical waste and excess water to the kidneys where it is collected and sent to the urinary bladder until the body excretes it as urine.

The urinary system is made up of a delicate system of organs, tubes, muscles, and nerves that work together to create, store, and carry urine. The urinary system includes two kidneys, two ureters, the urinary bladder, two sphincter muscles, and the urethra.

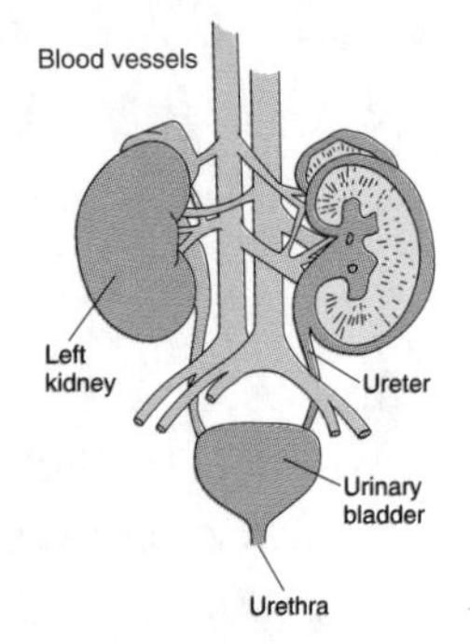

Your body takes nutrients from food and uses them to maintain all bodily functions, create energy, and repair damaged tissue in the body. After your body has taken what it needs from the food, waste products are left behind in the blood and in the bowel. The urinary system works with the lungs, skin, and intestines—all of which also excrete wastes—to keep the chemicals and water in your body balanced. The urinary system removes a type of waste called **urea** from your blood. Urea is produced when foods containing protein, such as meat, poultry, and certain vegetables, are broken down in the body. Urea is carried in the bloodstream to the kidneys.

The kidneys are bean-shaped organs about the size of your fists. They are near the middle of the back, just below the rib cage. The kidneys remove urea from the blood through tiny filtering units called **nephrons**. Urea, together with water and other waste substances, forms the urine as it passes through the nephrons and down the tubes of the kidney.

From the kidneys, urine travels down two thin tubes called ureters to the bladder. The bladder is a hollow muscular organ shaped like a balloon. The bladder stores urine until you are ready to go to the bathroom to empty it. It swells into a round shape when it is full and gets smaller when it is empty. Circular muscles called sphincters help keep urine from leaking. The sphincter muscles close tightly like a rubber band around the opening of the bladder into the urethra, the tube that allows urine to pass outside the body.

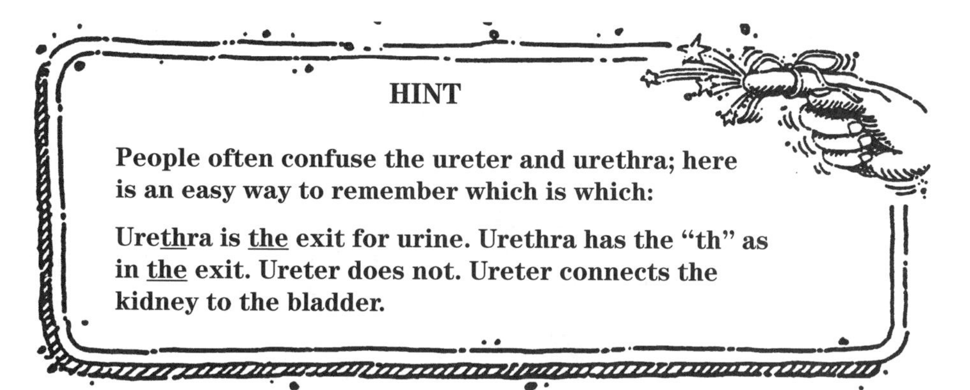

HINT

People often confuse the ureter and urethra; here is an easy way to remember which is which:

Urethra is the exit for urine. Urethra has the "th" as in the exit. Ureter does not. Ureter connects the kidney to the bladder.

Nerves in the bladder tell you when it is time to urinate, or empty your bladder. When you urinate, the brain signals the bladder muscles to tighten, squeezing urine out of the bladder. At the same time, the brain signals the sphincter muscles to relax. As these muscles relax, urine exits the bladder through the urethra.

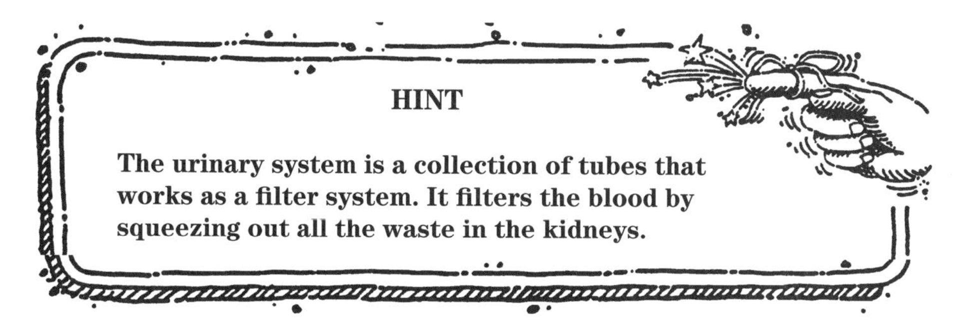

HINT

The urinary system is a collection of tubes that works as a filter system. It filters the blood by squeezing out all the waste in the kidneys.

LOCOMOTION SYSTEM

We walk, run, sit, lift a glass, and turn the pages of this book without thinking about how it happens. Each of these movements is the result of your body's **locomotion system** working together. All animals need to move around to get shelter and food. It is done by moving muscles and bones. The main structure of the locomotion system is the **skeleton**. It is made of more than 200 bones. Calcium and other minerals make the bone strong but slightly flexible. **Bone** is a living tissue with a blood supply; it is constantly being dissolved and repaired, especially when the bone is broken.

The skeleton has three main functions: support, protection, and movement.

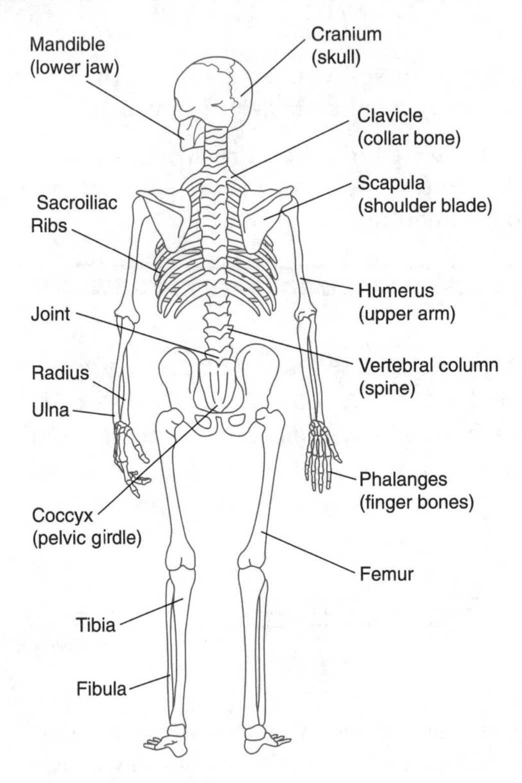

Support

The skeleton supports the body. For example, without a backbone, we would not be able to stay upright.

Protection

Here are some examples of what the skeleton protects:

- The skull protects the brain
- The rib cage protects the heart and lungs
- The backbone protects the spinal cord

Movement

Some bones in the skeleton are joined rigidly together and cannot move against each other. Bones in the skull are also joined like this. Other bones are joined to each other by flexible joints. **Muscles** are needed to move bones attached by joints.

If two bones just moved against each other, they would eventually wear away. This can happen in people who have arthritis. To keep this from happening, the ends of the bones in a joint are covered with a tough, smooth substance called **cartilage**. This is kept slippery by fluid found in between the cartilage. Tough **ligaments** join the two bones in the joint and stop it from falling apart. The diagram shows the main features of a **joint**.

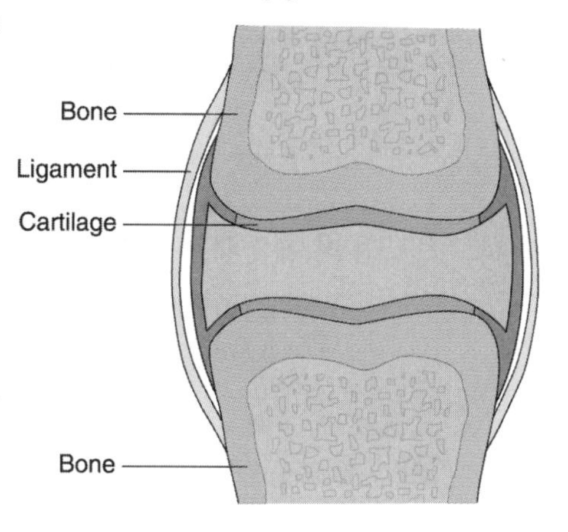

Different types of joints allow for different types of movement.

- Hinge joints allow simple movement, the same as a door opening and closing. Knee and elbow joints are hinge joints.
- Ball and socket joints allow movement in more directions. Hip and shoulder joints are ball and socket joints.

Bones can't move on their own, they need **muscle**s to make this happen. Muscles work by getting shorter—they **contract**; the process is called **contraction**.

Muscles are attached to bones by strong **tendons**. When a muscle contracts, it pulls on the bone, and the bone can move if it is part of a joint.

Muscles can only pull; they cannot push. This would be a problem if a joint was controlled by just one muscle. As soon as the muscle had contracted and pulled on a bone, that would be it, with no way to move the bone back again. The problem is solved by having muscles work in pairs.

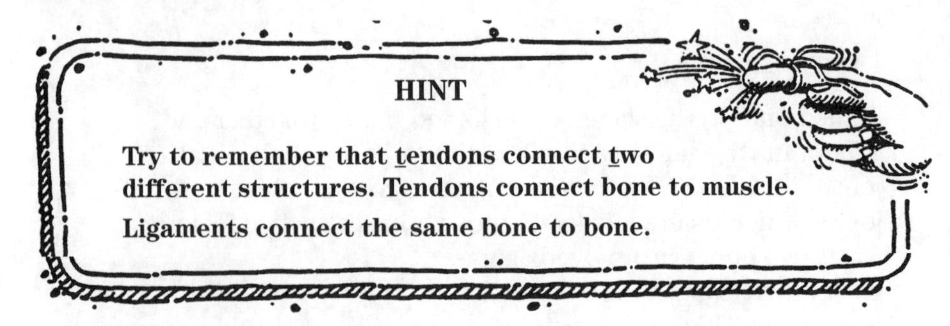

The elbow joint lets the forearm move up or down. It is controlled by two muscles: the **biceps** on the front of the upper arm and the **triceps** on the back of the upper arm.

- When the biceps muscle contracts, the forearm moves up.
- When the triceps muscle contracts, the forearm moves down.

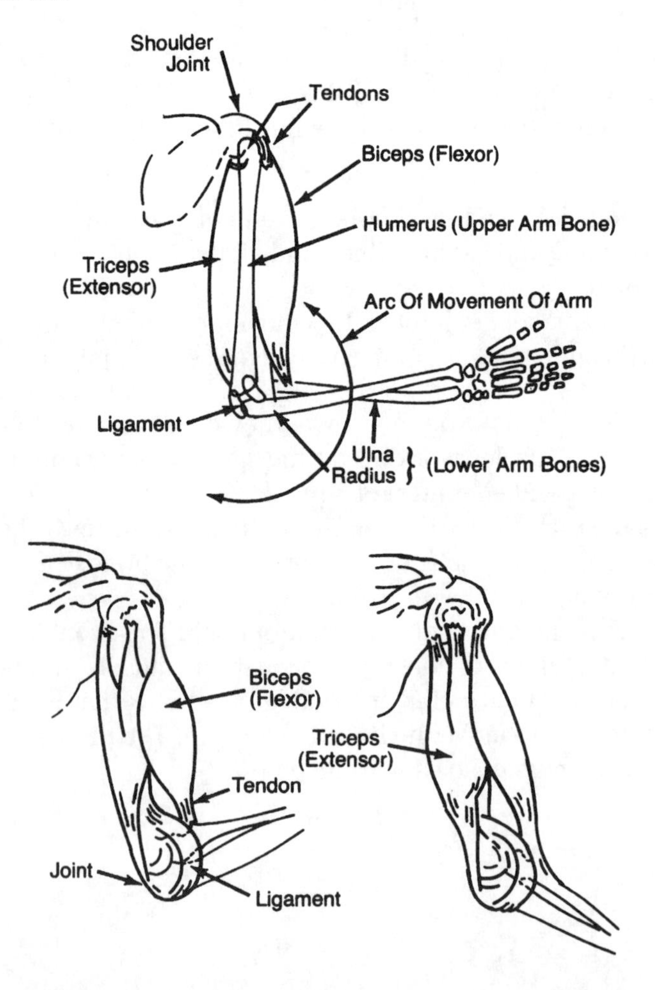

This solves the problem. To lift the forearm, the biceps contracts and the triceps relaxes. To lower the forearm again, the triceps contracts and the biceps relaxes.

Long bones, such as those in your arm and leg, are made up of bone marrow, surrounded by compact bone. We saw this earlier when we discussed where blood was produced.

BRAIN TICKLERS
Set # 27

1. What is the function of the urinary system?
2. Which organ of the urinary system is where the blood is filtered?
3. Label the urinary system.

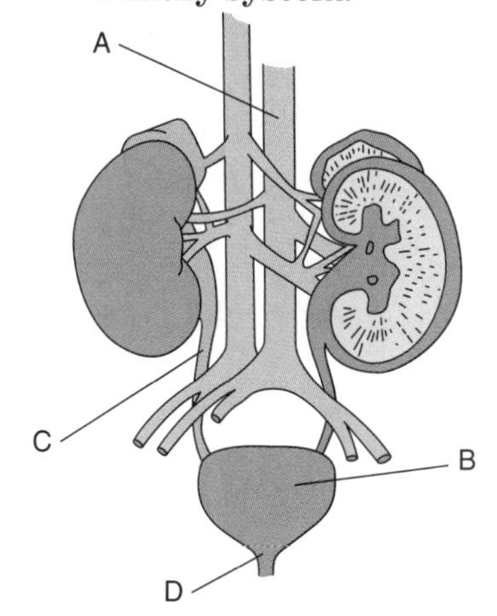

4. What information can be gained from taking a urine sample?
5. What is the function of the locomotion system?
6. Which structure attaches bone to bone?
   a. muscle
   b. tendon
   c. ligament
   d. cartilage

7. Which structure cushions joints?
   a. muscle
   b. tendon
   c. ligament
   d. cartilage
8. Which bone protects the brain?
9. Which bones protect the spinal cord?
10. Name two joints in your body.

(Answers are on page 157.)

NERVOUS SYSTEM

Walking, running, jumping, riding a bicycle, reading, eating, and talking are all things you do without thinking; they wouldn't be possible without your **nervous system**. Your nervous system is the control center of all that you do and think. The nervous system is made up of the **brain**, **spinal cord**, **nerves**, and **sense organs** such as your eyes and ears. Your nervous system carries information from your body to the brain and back to the body.

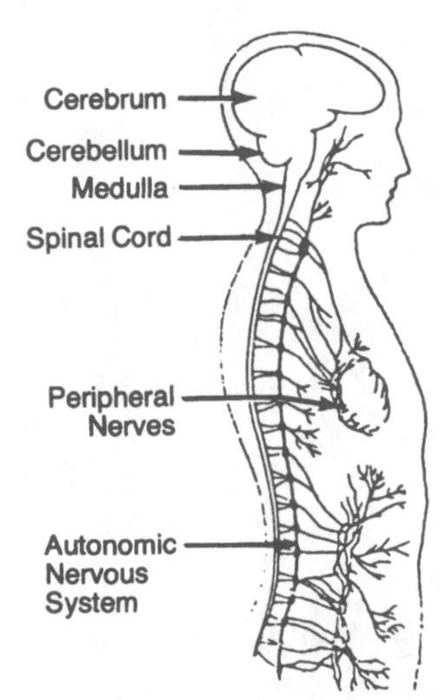

Information travels around the body along paths called nerves. The nerves are made up of cells that carry information through your nervous system; these cells are called are neurons. The messages that a neuron carries are called nerve **impulses**.

Neurons have a large body containing the nucleus, thread-like extensions called dendrites, and an axon.

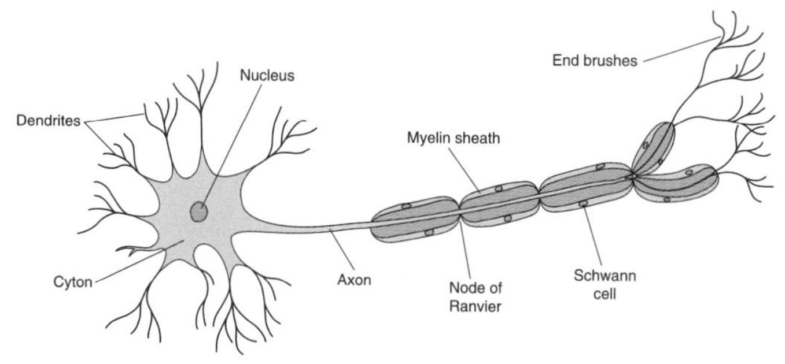

The **dendrites** carry impulses toward the neuron's cell body, while the **axon** carries impulses away. Nerve impulses begin in the dendrite and go from there to the cell body and on to the axon. A neuron has only one axon, but it can have many dendrites. There can, however, be more than one tip on the axon. Another name for the axon and dendrites are nerve fibers. Nerve fibers are arranged in parallel bundles and are covered with connective tissue. A bundle of nerve fibers is called a **nerve**.

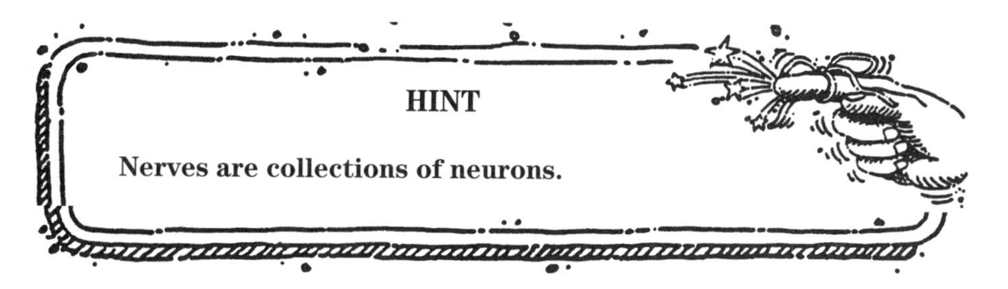

HINT

Nerves are collections of neurons.

Each nerve impulse begins in the dendrites of a neuron, which pick up stimuli from the area that the dendrites are in. The nerve impulse moves toward the cell body down the axon to the tip. It then travels along with the neuron in the form of electrical and chemical signals. When the impulse reaches the end of the neuron, it can pass to another neuron or it can travel to a muscle or organ. This point is called a **synapse**. When a nerve impulse passes at a synapse the dendrite and axon tip do not touch. The axon tips release chemicals that carry the impulse across.

There are three kinds of neurons in your body: sensory neurons, interneuron, and motor neurons. The nerve impulse travels in a path through these neurons from a stimulus to the brain and then to the reaction; the path is called the **reflex arc**.

If you put your hand on a something hot, your brain receives an impulse from the hand that you touched something hot, the brain tells the muscles of the arm and hand to pull your hand away from the hot object.

A **stimuli** is an action that influences an activity; it could be a sound, taste, touch, or sight. The **sensory neuron** picks up the stimuli and converts it into a nerve impulse; it senses the stimuli. The sensory neuron brings the impulse to the central nervous system, brain, and spinal cord. At the central nervous system the impulse jumps to the **interneuron**, which interprets the stimuli and decides if it is good or bad and if it needs a reaction. The impulse then moves to the **motor neuron**, which causes a reaction, usually a movement. A response could be a muscle movement, an increased heartbeat, a smile, a scream, or a cheer.

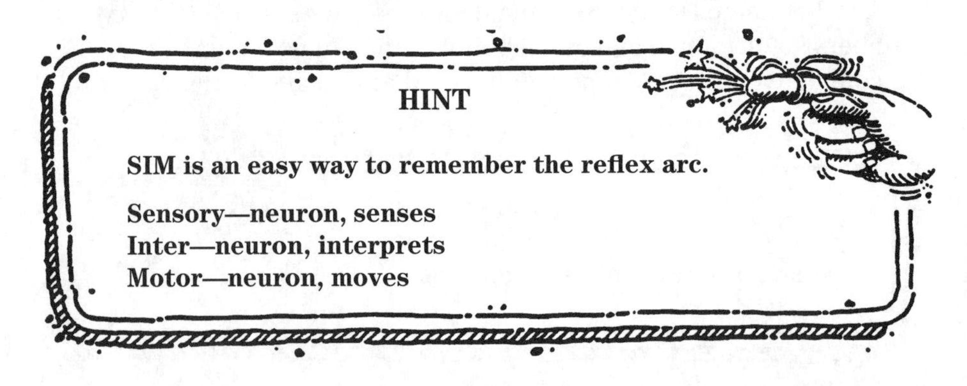

HINT

SIM is an easy way to remember the reflex arc.

Sensory—neuron, senses
Inter—neuron, interprets
Motor—neuron, moves

Anything that makes an organism react is called a stimulus. When an organism reacts to a stimulus it is called a response. There are two kinds of responses: a voluntary response and an involuntary response. Voluntary responses occur when you control the response such as swatting a fly. Your heartbeat however is involuntary.

Your nervous system is made up of two divisions, the **central nervous system** and the **peripheral nervous system**. The central nervous system is made up of the brain and spinal cord; the peripheral nervous system is made up of everything else.

The central nervous system

The main parts of the central nervous system are the brain and the spinal cord. The brain, which is located in your head, controls most of the functions of the body. The spinal cord is a very thick column of nervous tissue that links your brain to a lot of the nerves in the peripheral nervous system.

Your brain contains about 100 billion interneurons; these neurons can receive messages and give messages throughout the body. There are three main regions in the brain that process information: the cerebrum, the cerebellum, and the brain stem (medulla). The **cerebrum** is the largest part of the brain; different parts of the cerebrum control different movements, thoughts, speech, memories, and other brain functions that we control. The **cerebellum** is responsible for coordinating your muscles and helps keep your body moving. The impulses that tell your feet to walk start in the cerebellum. The **brain stem** (also known as the medulla), which lies between the cerebellum and the spinal cord, controls your involuntary actions such as breathing and heart rate. The **spinal cord** is the link between your brain and the peripheral nervous system. The spinal cord runs from the brain down the back and is protected by bones that surround it called vertebrae. Interneurons are part of the central nervous system.

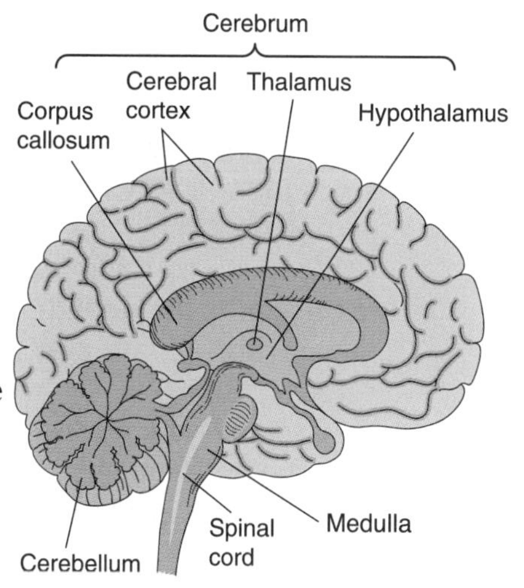

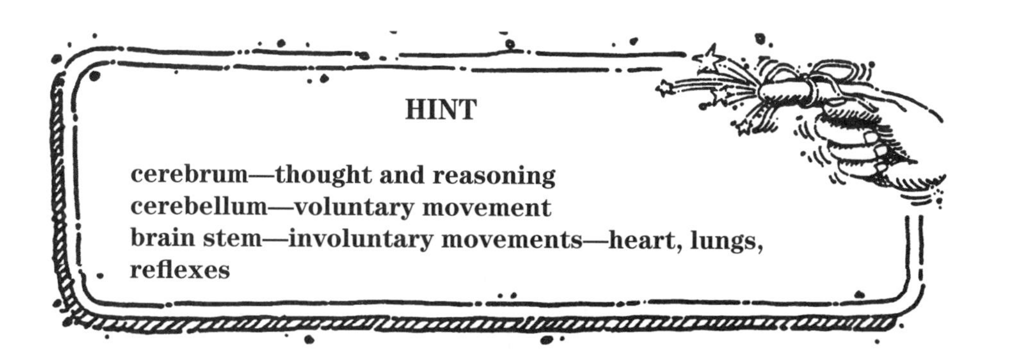

HINT

cerebrum—thought and reasoning
cerebellum—voluntary movement
brain stem—involuntary movements—heart, lungs, reflexes

The peripheral nervous system

The peripheral nervous system contains the network of nerves that branch out from the spinal cord and connect it to the rest of the body. The peripheral nerves are not protected by bone. These nerves spread to all of your organs, muscles, and limbs to help you do what you need to do each day. The sensory and motor neurons are part of the peripheral nervous system.

Humans have five senses: hearing, tasting, seeing, touching, and smelling. The senses tell you what is happening all around. These senses are interpreted in different areas of the brain.

We hear with our ears. The vibrations from sound move tiny bones in our ears, which send a message through the auditory nerve to the brain so that we understand what is being heard.

We taste with our tongues. The tiny sensors on the tongue are called taste buds; they can identify different tastes such as sweet, sour, bitter, and tart. Different areas of your tongue interpret different tastes. The taste buds send the information to the brain so that we can figure out if we like or dislike the taste.

We see with our eyes. The eyes take in light through the lenses and transmit the shapes and changes through the optic nerve to the brain.

We touch with our fingers and skin. All over the skin there are tiny spots that feel things and transmit the texture, temperature, and other characteristics to the brain by way of the nerves.

We smell with our noses. Tiny hairs and nerves in our nose sense the different smells and send them to the brain so that we can decide if it is a familiar smell or a new one.

BRAIN TICKLERS
Set # 28

1. The cell of the nervous system is the __________.
2. Your finger gets a pin stuck in it; trace the path of the nerve impulse.
3. The brain and spinal cord are part of the __________ nervous system.
4. Hearing and vision are considered __________.
5. Which part of the brain controls voluntary movement?
   a. cerebellum
   b. cerebrum
   c. brain stem
   d. spinal cord
6. Which part of the brain controls involuntary movement?
   a. cerebellum
   b. cerebrum
   c. brain stem
   d. spinal cord
7. A bundle of neurons is called a(n) __________.

(Answers are on page 158.)

ENDOCRINE SYSTEM

The **endocrine system** regulates how we grow, develop, go through puberty, reproduce, and process foods (also known as our metabolism); it also plays a part in determining our mood. The endocrine system is made up of **glands** that are found all over the body. The glands secrete **hormones** that cause changes in organs in the body. The glands of the endocrine system and the hormones they release influence almost every cell, organ, and function of our bodies.

Simply stated, the endocrine system is in charge of body processes that happen slowly, such as cell growth. Faster processes like breathing and body movement are controlled by the nervous system. Even though the nervous system and endocrine system are separate systems, they often work together to help the body function properly.

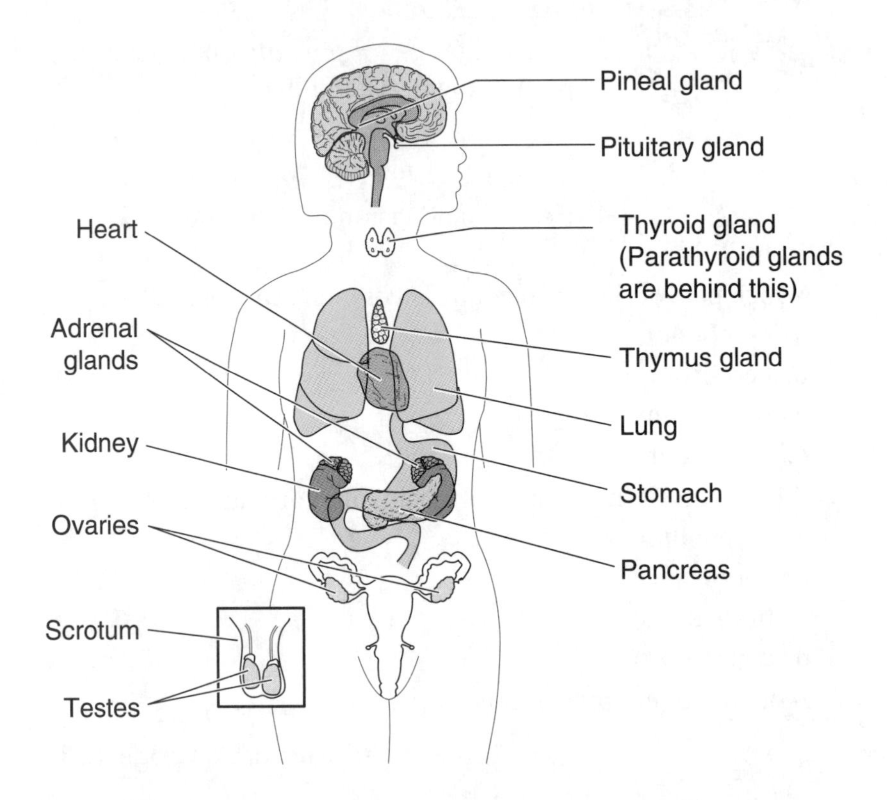

Hormones are the body's **chemical messengers**; hormones transfer information and instructions from one set of cells to another. Although many different hormones circulate throughout the bloodstream, each one affects only the cells that are genetically programmed to receive and respond to its message. Hormone levels can be influenced by factors such as stress, infection, and changes in the balance of fluids and minerals in the blood.

HINT

Our body has two types of glands: endocrine and exocrine.

endocrine glands—Secretions go into the blood to cause a change on a different area of the body called the target.

exocrine glands—Secretions exit organs into areas. Examples are salivary glands and gastric glands.

A gland is a group of cells that produces and secretes, or gives off, chemicals. Some types of glands release their secretions in specific areas; others secrete into the bloodstream where they can be transported to cells in other parts of the body.

The major glands that make up the human endocrine system are the hypothalamus, pituitary, thyroid, parathyroid, adrenals, pineal gland, and the reproductive glands, which include the ovaries and testes. The pancreas is also part of this hormone-secreting system, even though it is also associated with the digestive system because it produces and secretes digestive enzymes.

The **hypothalamus** is a collection of specialized cells that is located in the lower central part of the brain. It is the primary link between the endocrine and nervous systems. Nerve cells in the hypothalamus control the pituitary gland by producing chemicals that either stimulate (speed up) or suppress (slow down) hormone secretions from the pituitary.

Although it is no bigger than a pea, the **pituitary gland** is considered the most important part of the endocrine system. It's often called the "master gland" because it makes hormones that control several other endocrine glands. The pituitary gland secretes hormones that are responsible for growth, metabolism, milk production, reproduction, and heart rate.

The **thyroid**, located in the front part of the lower neck, is shaped like a bowtie or butterfly and produces the thyroid hormone **thyroxine**. This hormone controls the rate at which cells burn fuels from food to produce energy. As the level of thyroid hormones increases in the bloodstream, the chemical reactions in the body speed up. Thyroid hormones also play a key role in bone growth and the development of the brain and nervous system in children.

Attached to the thyroid are four tiny glands that function together called the **parathyroids**. They release parathyroid hormone, which regulates the level of calcium in the blood, which in turn helps in the development of strong bones.

The body has two adrenal glands, one on top of each kidney. The **adrenal glands** secrete hormones that influence or regulate salt and water balance in the body, the body's response to stress, metabolism, the immune system, and sexual development and function. They also secrete a hormone called **adrenaline** (also known as epinephrine), which increases blood pressure and heart rate when the body experiences stress.

The **pineal gland**, is located in the middle of the brain. It secretes **melatonin**, a hormone that may help regulate the wake–sleep cycle.

The **gonads** are the main source of sex hormones. Male gonads, **testes**, are located in the scrotum and secrete the hormone **testosterone**. This hormone regulates body changes during puberty and the appearance of other male secondary sex characteristics such as the deepening of the voice, growth of facial and pubic hair, and the increase in muscle growth and strength. Working with hormones from the pituitary gland, testosterone also supports the production of sperm by the testes.

The female gonads, the **ovaries**, are located in the pelvis. They produce eggs and secrete the female hormone **estrogen** and **progesterone**. Estrogen is involved in the development of female sexual features such as breast growth, the accumulation of body fat around the hips and thighs, and the growth spurt that occurs during puberty. Both estrogen and progesterone are also involved in pregnancy and the regulation of the menstrual cycle.

The **pancreas** produces two important hormones, **insulin** and **glucagon** (in addition to others). They work together to maintain a steady level of glucose, or sugar, in the blood and to keep the body supplied with fuel to produce and maintain stores of energy.

The endocrine system works by sending chemical messages. Once a hormone is secreted, it travels from the endocrine gland through the bloodstream to target cells designed to receive its message.

For example, if the thyroid gland has secreted adequate amounts of thyroid hormones into the blood, the pituitary gland senses the normal levels of thyroid hormone in the bloodstream and adjusts its release of thyrotropin, the pituitary hormone that stimulates the thyroid gland to produce thyroid hormones.

REPRODUCTION/LIFE CYCLES

Humans go through different stages in their lives. They are born, grow (develop), have babies (offspring), and die. This is called the continuity of life or **life cycle**. The process by which a baby is created and born is called **reproduction**. The offsprings grow and change over time. This is called **development**.

Humans reproduce sexually. During sexual reproduction, an **embryo** is formed when an egg carried by the female is fertilized by the male. When the two cells join, they merge two sets of genetic information, one from the father and one from the mother, to make a baby that is genetically different from either parent.

Each reproductive system contains the organs needed for reproduction or producing babies. Let's learn about the male and female reproductive systems.

Fertilization and fetal development

Fertilization happens when an egg cell joins with a sperm cell to create a fertilized egg. Humans have internal fertilization, sperm cells travel from the male reproductive system into the female reproductive system. When a sperm cell meets with an egg cell, fertilization can happen.

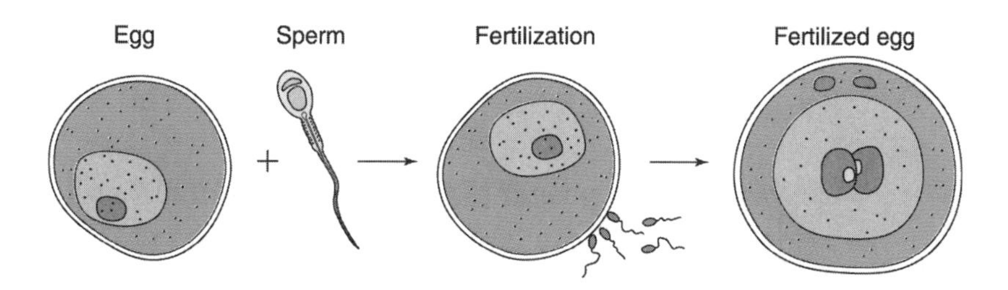

The fertilized egg divides to form a ball of cells called an **embryo**. This attaches to the lining of the uterus and begins to develop into a **fetus** and finally a baby.

The fetus relies upon its mother as it develops. Inside the mother the fetus is protected, gets oxygen and nutrients, and gets rid of wastes. The fetus is protected by the uterus and the **amniotic fluid**, a liquid contained in a bag called the **amnion**.

The placenta is responsible for providing oxygen and nutrients and removing waste substances. It grows into the wall of the uterus and is joined to the fetus by the **umbilical cord**. The mother's blood does not mix with the fetus's blood, but the placenta lets substances pass between the two blood supplies:

- Oxygen and nutrients **diffuse** across the placenta from the mother to the fetus
- Waste substances, such as carbon dioxide, diffuse across the placenta from the fetus to the mother

Throughout the nine months of a pregnancy, the fetus goes through a series of changes. It grows and differentiates from a single cell into a ball of cells and then into an oval-looking shape. It sprouts arm and leg buds and then develops into the shape of a baby. During this time, cells divide and tissues, organs, and organ systems form, so that the baby will be able to survive on its own once it is born.

After nine months the baby is ready to be born; this is called the **birth**.

BRAIN TICKLERS
Set # 29

1. The endocrine system secretes __________ that cause __________ in other parts of the body.
2. The master gland found in the brain is the __________.
3. The glands found on the kidneys are the __________; they regulate your heart rate.

4. The sex glands are called the __________; they are the __________ in the male and the __________ in the female.
5. What gland controls the sugar levels in your body?
6. What is the function of the reproductive system?
7. The joining of an egg and sperm is called __________ and takes place inside the (*circle one:* male or female) body.
8. Humans reproduce by __________ reproduction.

(Answers are on page 158.)

WRAPPING UP

- The human body works like a machine. Cells make up tissues, tissues make up organs, and organs make up organ systems. Each organ system has a specific function that helps the human perform his or her life function.
- The digestive system consists of three phases: ingestion, digestion, and egestion. Food enters the mouth and passes through the body where it is broken down mechanically and chemically. Nutrients are removed from the food and distributed throughout the body once it is absorbed into the bloodstream in the small intestine.
- The circulatory system is a closed system of tubes that pump blood to every cell in the body. Veins carry blood to the heart; the heart pumps blood to the lungs and back so that it can exchange carbon dioxide for oxygen. The blood then goes into the arteries, and finally exchange occurs between cells at the capillaries.
- Metabolic wastes leave the body through the urinary system. The kidneys filter the metabolic waste from the blood and store it in the bladder until urination.
- The body moves a system of bones, muscles, cartilage, ligaments, and tendons, which make up the locomotion system.
- The body is controlled by the nervous system. The nervous system is made up of specialized cells called neurons that interpret stimuli all over the body with the help of the brain and spinal cord.

- The endocrine system regulates how humans grow, develop, reproduce, process food, and go through other life functions. It does this by having glands that secrete hormones into the blood. Hormones trigger various responses in the body.
- The human reproductive system is specialized so that a genetically different offspring can result from sperm and egg joining inside the female body.

BRAIN TICKLERS—THE ANSWERS

Set # 25, page 131

1. The digestive system breaks down nutrients for absorption into the bloodstream for use by the cells of the body.
2. mouth
3. bile; small intestine
4. small intestine
5. b. liver
6. a. esophagus
7. The respiratory system takes oxygen into the body for use by the cells and removes the gaseous waste, carbon dioxide, from the body (gas exchange).
8. nasal cavity → trachea → bronchi → bronchiole → alveoli → gas exchange
9. Smoking will interfere with gas exchange in the alveoli and can damage the cells of the alveoli.
10. In the nasal cavity the hair filters the air, the blood warms the air, and the mucus moistens the air entering the body.

Set # 26, page 137

1. aorta → artery → capillary → veins → vena cava
2. The right side of the heart has oxygen-poor (deoxygenated) blood; the left side carries oxygen-rich blood.

3. The circulatory system transports blood throughout the body. The blood carries nutrients to the cells of the body and wastes from the cells out of the body.
4. b. red blood cells
5. d. plasma
6. bone marrow
7. four; A, B, AB, and O.
8. The skin, respiratory surfaces in the nose and esophagus and mouth and stomach provide a first barrier to eliminate a pathogen before it enters the body's bloodstream.
9. Vaccines provide the body with a dead strain of a disease so that the body can fight the disease if it gets it. This is called acquired immunity or active immunity.
10. A, right atrium; B, right ventricle; C, left atrium; D, left ventricle

Set # 27, page 143

1. The urinary system filters chemical wastes from the blood.
2. kidney
3. A, kidneys; B, bladder; C, ureter; D, urethra
4. A urine sample shows all of the materials that the body has processed, if it has diseases, if there are infections, as well as any chemicals in the body. These appear during a urine test.
5. protection, support, and movement
6. b. tendon
7. d. cartilage
8. skull
9. vertebrae
10. Answers may vary. Joints include knee, ankle, elbow, shoulder, and wrist.

Set # 28, page 149

1. neuron
2. When your finger gets a pin stuck in it, sensory neurons in finger and arm send a message to the interneuron in the spinal cord. That is the stimulus. The motor neuron tells the muscle to pull out the pin and the mouth to yell "Ouch!" That is the response.
3. central
4. senses
5. a. cerebellum
6. c. brain stem
7. nerve

Set # 29, page 154

1. hormones; changes
2. pituitary
3. adrenal
4. gonads; testes; ovaries
5. pancreas
6. The reproductive system is responsible for producing sex cells and joining them together during fertilization to form an offspring.
7. fertilization, female
8. sexual

Genetics

Have you ever noticed that some children look like their parents and some may even look like their grandparents? How does this happen?

GENETICS

Genetics is the branch of science that studies how hereditary information is passed on from one generation to the next generation in a family.

Why do you look the way you do? Our characteristics are often referred to as **traits**. A trait could be the color of your hair, skin, or eyes; your height; or the shape of your face, ears, or nose. These traits come from our **genes**. A gene is the basic physical and functional unit of heredity. Genes, are made up of DNA; they act as instructions to make you what you are. In humans, genes vary in size from a few hundred DNA bases to more than two million bases. A **base** is a chemical compound that holds the DNA components. Every human has two copies of each gene, one inherited from each parent. Genes are responsible for various characteristics such as height, eye color, shape of nose, shape of ear, and blood type. Variations of the gene relating to the same trait are called **alleles**. Individuals carry two genes for each trait, one from the mother's egg and one from the father's sperm.

If an organism is the result of asexual reproduction, the offspring will have the exact same genetic information as the parent. If an organism is the result of sexual reproduction, the offspring will get genetic information from both parents and be genetically different from the parents.

DNA (deoxyribonucleic acid) is a bundle of chemicals that are considered the map for all living things. From the nucleus of cells they can control your body. Everyone's DNA is unique. Not only do you have a unique fingerprint, but you also have your own DNA.

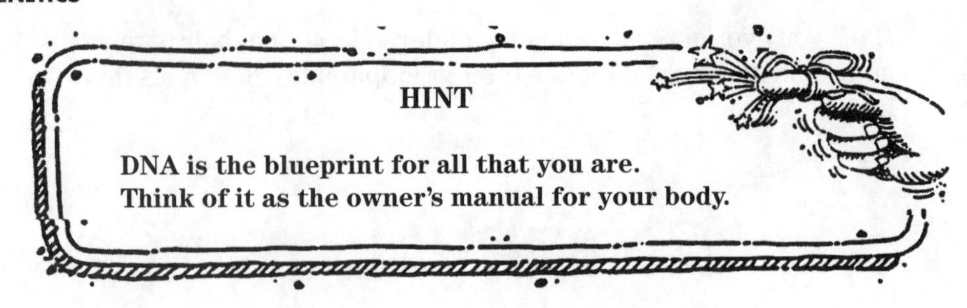

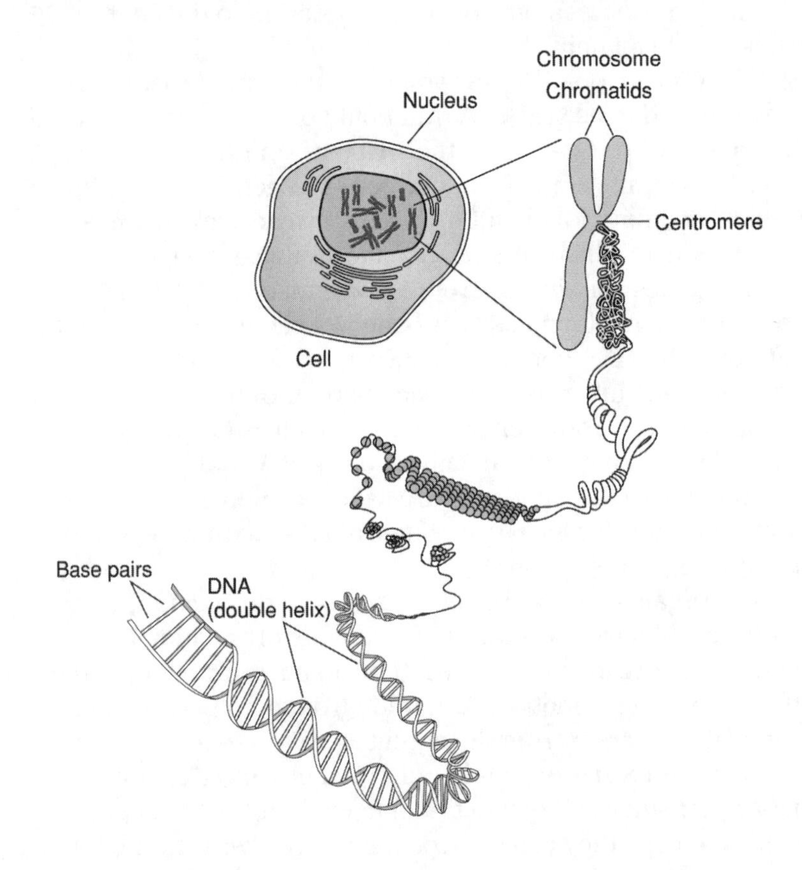

What is a chromosome?

In the nucleus of each cell, the DNA molecule is packaged into thread-like structures called **chromosomes**. Each chromosome is made up of DNA tightly coiled many times around proteins. If the chromosomes in one of your cells were uncoiled and placed end to end, the DNA would be about 6 feet long. If all of the DNA in your body were connected in this way, it would stretch approximately sixty-seven billion miles! That's nearly 150,000 round trips to the moon.

Chromosomes are not visible in the cell's nucleus—not even under a microscope—when the cell is not dividing. However, the DNA that makes up chromosomes becomes more tightly packed during cell division and is then visible under a microscope. Most of what researchers know about chromosomes was learned by observing chromosomes during cell division.

Each chromosome has a center point where the two single strands called **chromatids** meet; the point is called the **centromere**, which divides the chromosome into two sections.

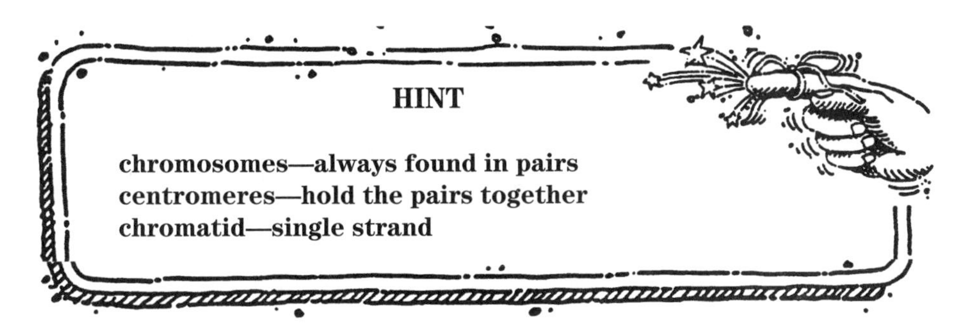

HINT

chromosomes—always found in pairs
centromeres—hold the pairs together
chromatid—single strand

Each species has a different number of chromosomes.

Species	#	Species	#
Common fruit fly	8	Human	46
Gorillas, Chimpanzees	48	Dog	78
Elephants	56	Goldfish	100–104

Every cell in your body except for eggs, sperm, and red blood cells contains a full set of chromosomes in its nucleus.

A **family tree** is a chart that shows one's relatives over time. In the family tree chart on the next page, the grandparents are the first generation, the parents are the second generation, and the baby would be the third generation.

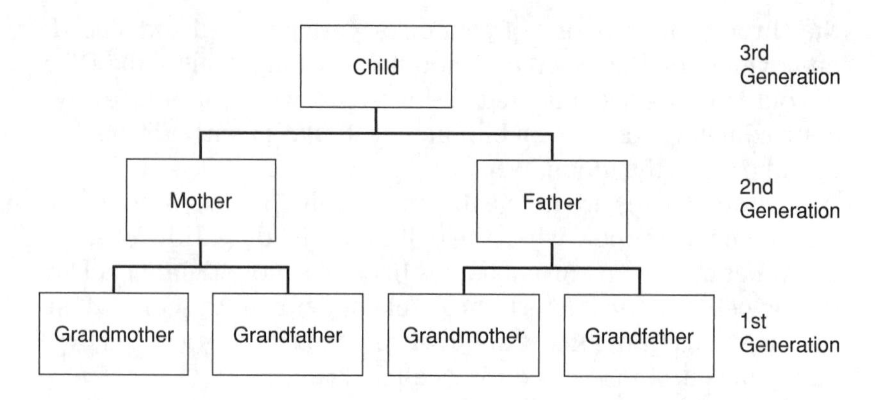

A pedigree chart shows each level of the family tree in a chart format. Lines and boxes connect parents and children from one level to the next.

Have you ever seen a large family where each child looks different from the others? Then there are other families where all of the children look alike. How does this happen? When organisms reproduce sexually, an equal amount of DNA from the mother and father combine to form the genetic information/DNA for the new offspring. This DNA provides the offspring with the information that defines the offspring's appearance. The new DNA is found in the nucleus of all of the offspring's cells.

mother's DNA + father's DNA = child's DNA

The man credited with the original theories of genetics was Austrian **Gregor Mendel** (1822–1884). Mendel is known as the father of genetics. Even though he did his research 100 years before the discovery of DNA, his theories are still studied and have been proven.

Mendel researched how pea plants pass on their traits from one generation of plant to the next. These **inherited traits** for a plant were the color of their flowers, the shape of their leaves, the shape and color of their seeds, and their overall height.

He chose pea plants to study because they reproduce quickly and he would be able to see many generations in a short period of time. Today geneticists study fruit flies because they have a very short life span. Because fruit flies live only a few weeks at the most, they are able to observe many generations of traits in a relatively short period of time.

So how do two dark-haired parents have a blond-haired child? Mendel explained this as a principle of **segregation**. The principle of segregation has three parts:

- Hereditary traits are determined by specific genes.
- Individuals carry two genes for each trait: one from the mother's egg and one from the father's sperm.
- When an individual reproduces, the two genes split up (segregate) and end up in separate gametes (sex cells).

BRAIN TICKLERS
Set # 30

1. Who is the father of genetics?
2. Explain why genetic researchers use fruit flies or pea plants to do their studies.
3. Different forms of a gene are called __________.
   a. alleles
   b. chromatid
   c. phenotype
   d. pedigree
4. The scientific study of heredity is called __________.
5. DNA is found in the __________.

(Answers are on page 179.)

HOW DOES DNA TRANSMIT INFORMATION?

The information in DNA is stored as a code made up of four chemical bases: **adenine (A)**, **guanine (G)**, **cytosine (C)**, and **thymine (T)**. The order, or sequence, of these bases determines the information available for building and maintaining an organism, similar to the way in which letters of the alphabet appear in a certain order to form words and sentences.

DNA bases pair up with each other, A with T and C with G, to form units called **bases**. Each base is also attached to a sugar molecule and a phosphate molecule. Together, a base, sugar, and phosphate are called a **nucleotide**. Nucleotides are arranged in two long strands that form a spiral called a double helix. The structure of the double helix is somewhat like a ladder, with the base pairs forming the ladder's rungs and the sugar and phosphate molecules forming the vertical sidepieces of the ladder.

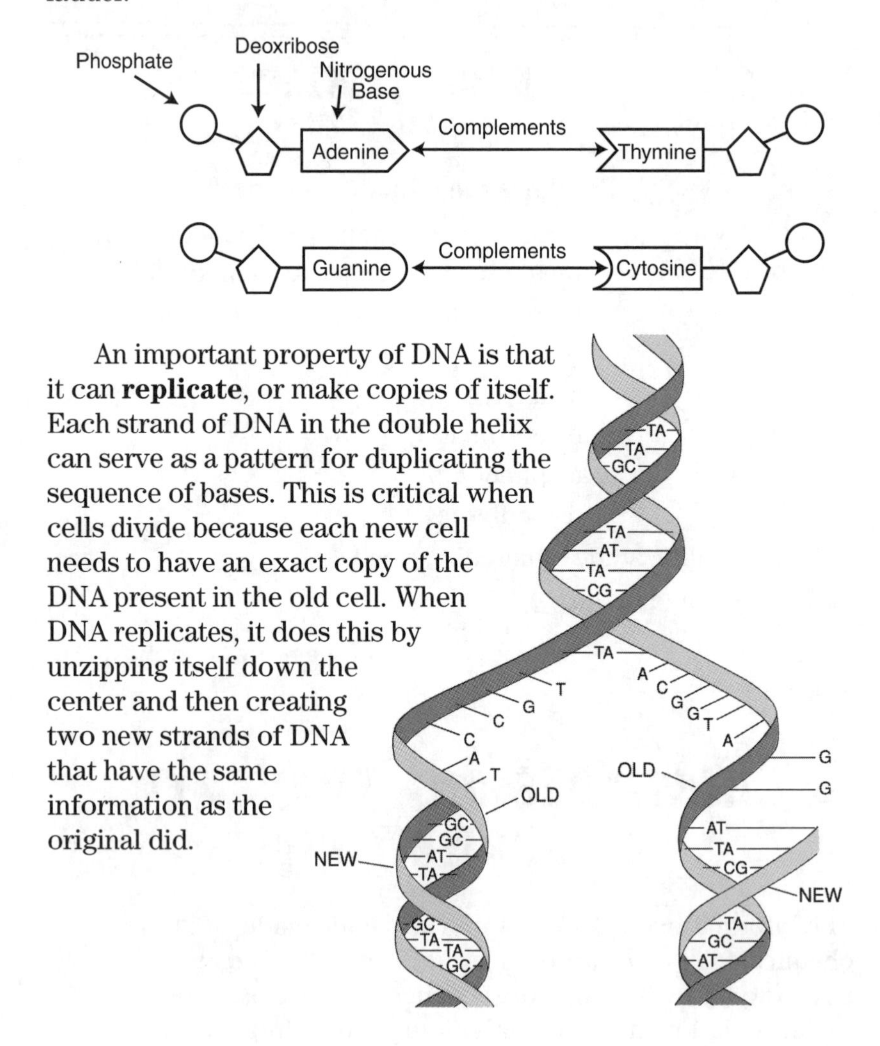

An important property of DNA is that it can **replicate**, or make copies of itself. Each strand of DNA in the double helix can serve as a pattern for duplicating the sequence of bases. This is critical when cells divide because each new cell needs to have an exact copy of the DNA present in the old cell. When DNA replicates, it does this by unzipping itself down the center and then creating two new strands of DNA that have the same information as the original did.

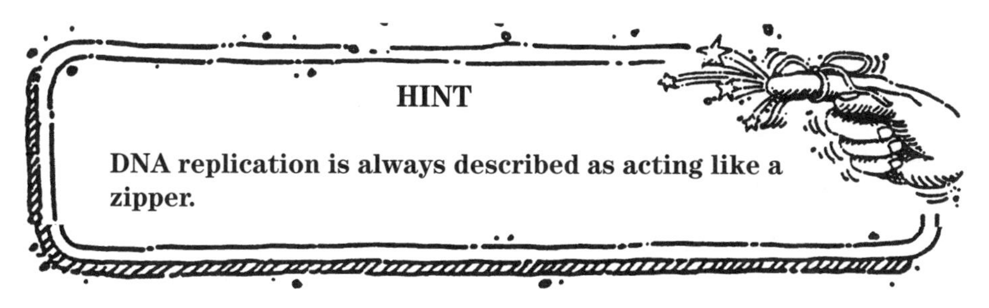

Most of your cells have what is called a **diploid** number of chromosomes because they have two sets of chromosomes (twenty-three pairs). Eggs and sperm are also known as sex cells or gametes; these are known as **haploid** cells. Each haploid cell has only one set of twenty-three chromosomes so that at fertilization the diploid number (46) is restored.

Chromosomes are numbered 1 to 22, according to size, with 1 being the largest chromosome. The twenty-third pair, known as the sex chromosomes, are called X and Y. Females have two copies of the **X chromosome**, while males have one X and one **Y chromosome**.

Male: XY   Female: XX

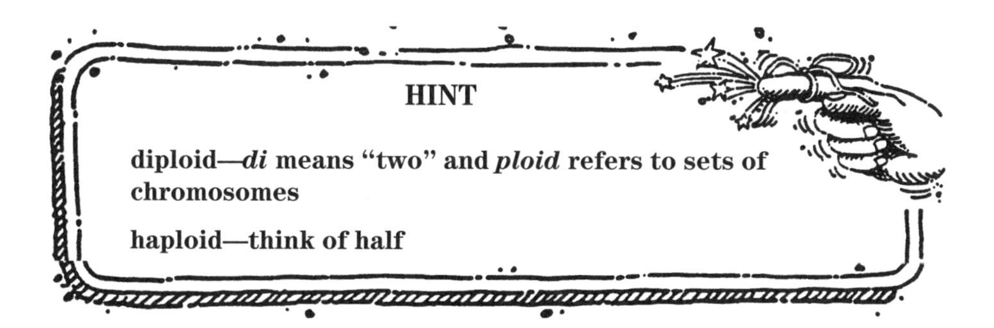

In humans, each cell normally contains twenty-three pairs of chromosomes, for a total of forty-six. Twenty-two of these pairs, called **autosomes**, look the same in both males and females. The twenty-third pair, the **sex chromosomes**, differs between males and females.

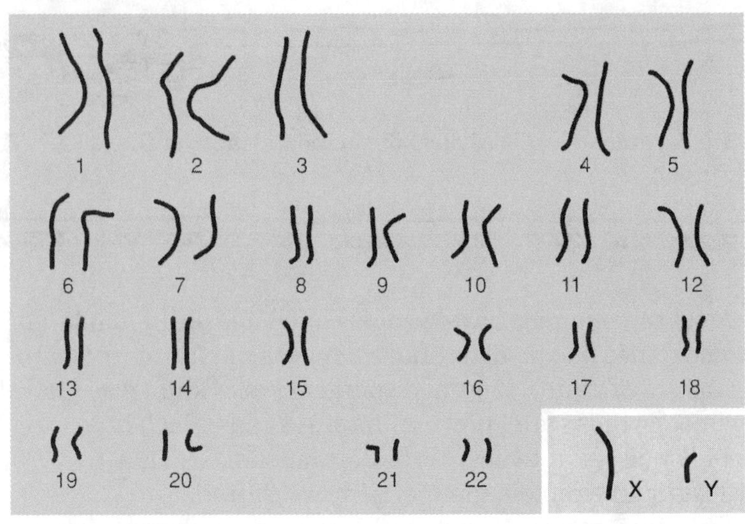

This picture of the human chromosomes lined up in pairs is called a **karyotype**. A karyotype is a photomicrograph of chromosomes arranged according to a standard classification. The twenty-two autosomes, body cells, are numbered by size. The other two chromosomes, X and Y, are the sex chromosome, which means this is from a male.

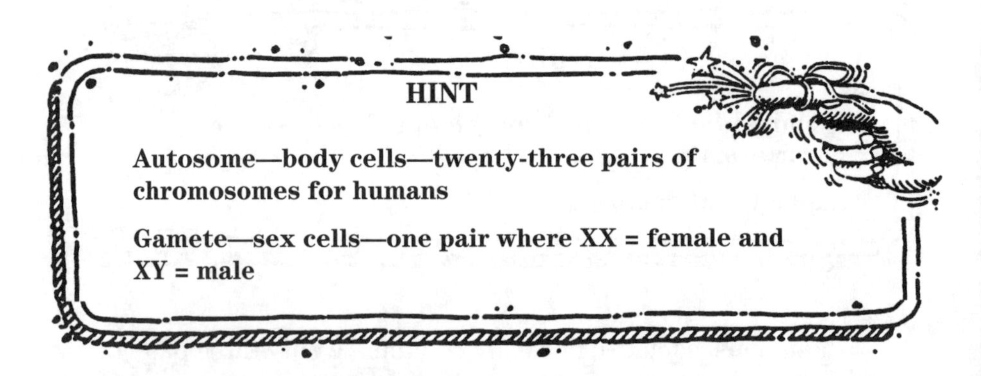

HINT

Autosome—body cells—twenty-three pairs of chromosomes for humans

Gamete—sex cells—one pair where XX = female and XY = male

When an individual cell reproduces, the two genes split up (segregate) and end up in different gametes. This is explained by the process called meiosis. **Meiosis** is like mitosis (normal cell division) but instead produces sex cells (gametes: sperm and egg).

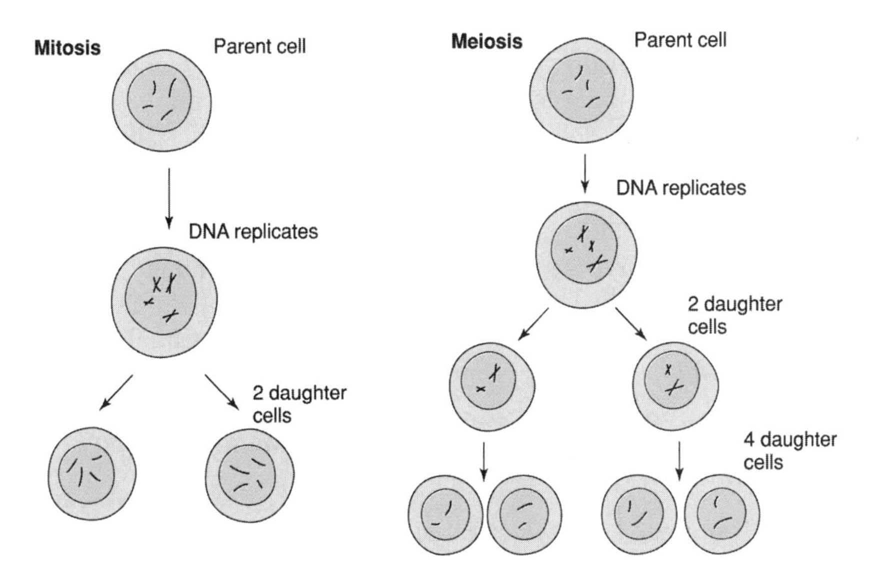

The sorting of genes occurs in the cell during meiosis. **Alterations** to the chromosomes' structure and organization can result in mutations that cause diseases.

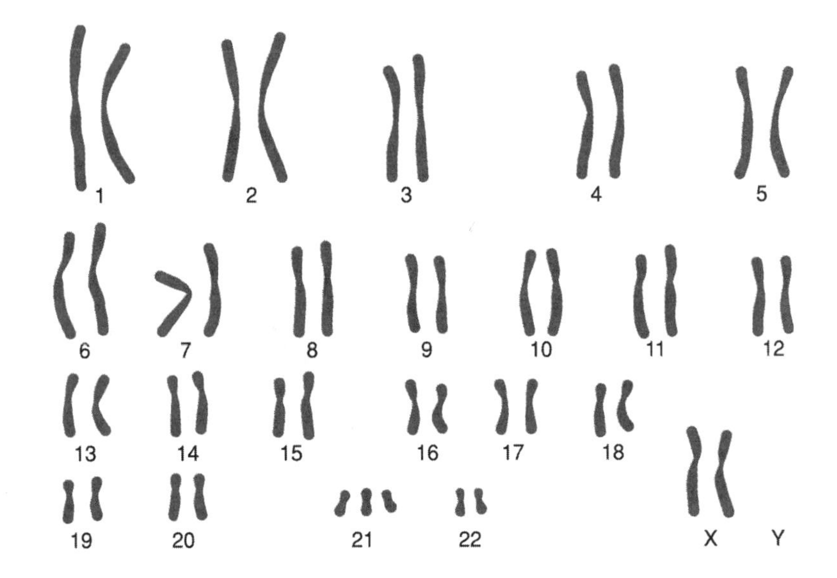

Trisomy 21, a genetic disorder that occurs when either the egg or sperm has an additional chromosome (# 21), is also known as **Down syndrome**. The child born with Down syndrome often has physical characteristics and other medical conditions that come from this additional chromosome. Other chromosomal disorders come about when there are fewer than forty-six chromosomes or even broken chromosomes.

Gene expression

Specific traits or alleles can be either **dominant**—strong, always appear or **recessive**—weak, easily hidden. The dominant allele will mask the recessive. For example, if the father gives a tall allele of the height gene and the mother gives a short allele, the offspring will be tall. This is because tall is dominant and short is recessive.

If we are to understand gene expression, we need to know these terms:

- **Genotype**—the genetic makeup, whether the alleles are dominant or recessive
- **Phenotype**—the physical appearance, what the person or organism looks like
- **Homozygous**—the same trait for both alleles, both dominant or both recessive
- **Heterozygous** or **hybrid**—different traits for the alleles, with one dominant and one recessive

Punnett square

A British mathematician/biologist named R.C. Punnett devised a method of picturing this concept on a graph called a **Punnett square**. Punnett squares graph the father's genotype (the genetic information concerned with a specific trait, for example, two alleles for tall, or two for short, or one for each) crossed with the mother's. Punnett squares show the **probability** of having children who have a certain trait. Probability explains why all of the children look alike in some families and so different in others.

- Dominant alleles are shown by a capital letter.
- Recessive alleles are shown by the lowercase of the same letter.

Alleles	Genotype	Phenotype
T	Dominant	Tall
t	Recessive	Short
TT	Homozygous Dominant	Tall
tt	Homozygous Recessive	Short
Tt	Hetorozygous or Hybrid	Tall

T = tall, t = short

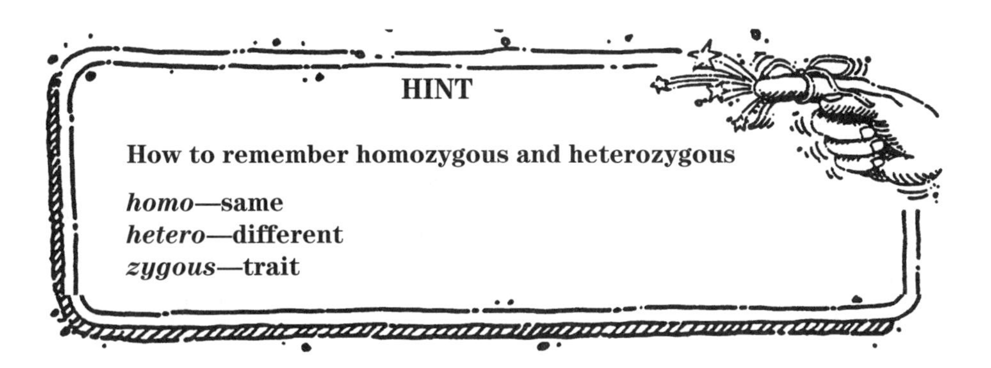

To do the Punnett square, match the top and side letters in each box.

This graph is a cross between a mother who is a hybrid or heterozygous for tall (meaning she has one allele (T) for tallness and one (t) for shortness) and a father who is homozygous short. Physically she is tall because T is dominant and masks the shortness genes from the father. This means that the parents have a 2/4 or 50 percent chance of having tall children and a 2/4 or 50 percent chance of having short children. This is a 1:1 ratio.

		Dad	
		t	t
Mom	T	**Tt**	**Tt**
	t	**tt**	**tt**

Based on this Punnett square, all kids will be tall, 4:0 ratio:

- 50 percent will be pure tall
- 50 percent will be hybrid tall

		Dad	
		T	T
Mom	T	**TT**	**TT**
	t	**Tt**	**Tt**

This 3:1 ratio of tall to short means that there is a

- 75 percent chance of being tall
- 25 percent chance of being pure tall
- 50 percent chance of being hybrid tall
- 25 percent chance of being short

		Dad	
		T	T
Mom	T	TT	Tt
	t	Tt	Tt

for every child that is conceived.

BRAIN TICKLERS
Set # 31

1. The letter that represent the base pairs for DNA are __________.
2. Describe the difference between genotype and phenotype.
3. What can be determined by this karyotype?

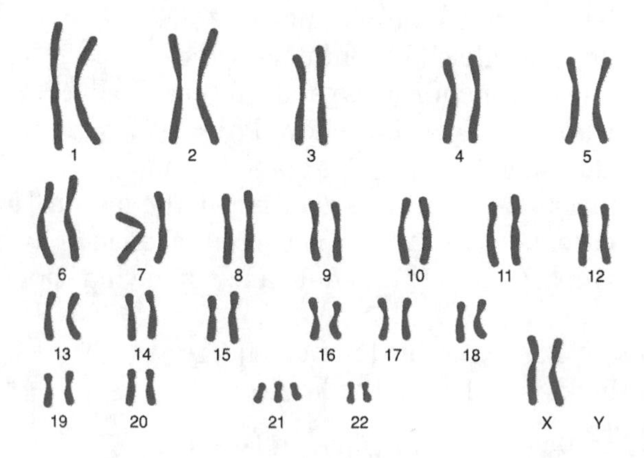

4. Draw a Punnett square for mating of a homozygous dominant TT parent and a homozygous recessive tt parent. TT × tt, calculate the probability for having homozygous dominant TT, homozygous recessive tt, and hybrid Tt children.

	T	T
t		
t		

(Answers are on page 179.)

BLOOD TYPING AND INHERITANCE

There are three alleles or versions of the **blood type** gene: A, B, and O. Since everybody has two copies of these genes, there are six possible combinations: AA, BB, OO, AB, AO, and BO. In genetic terms, these combinations are the genotypes children get from their parents.

In addition to the proteins (agglutinogens, clumping proteins) existing on your red blood cells, other genes make the proteins that circulate in your blood plasma. These proteins are responsible for ensuring that only the blood cells of your blood type exist in your body.

Your genotype determines your blood type. The protein produced by the O allele has no special enzymes. However, the protein produced by thc A and B alleles do have enzymatic activities, which are different from each other. Therefore people whose genotype is OO are said to have type O blood, meaning the protein on their red blood cells doesn't have any enzymatic activity. People with type O blood have agglutinins proteins A and B in their blood plasma. Agglutinin protein helps the body destroy any type A blood cells that might enter the circulation system. Agglutinin B helps the body destroy any type B blood cells that might enter the circulation system.

HINT

Think of blood type O as having no proteins, so it can't have any A, B, or AB proteins.

GENOME

Life is specified by **genomes**. Every organism, including humans, has a genome that contains all of the biological information needed to build and maintain a living example of that organism. The biological information contained in a genome is encoded in its DNA and is divided into genes. Genes code for proteins that attach to the genome at the appropriate positions and switch on a series of reactions called gene expression.

The **Human Genome Project** was an international research effort to determine the sequence of the human genome and identify the genes that it contains. The project was coordinated by the National Institutes of Health and the U.S. Department of Energy. Additional contributors included universities across the United States and international partners in the United Kingdom, France, Germany, Japan, and China. The Human Genome Project formally began in 1990 and was completed in 2003, two years ahead of its original schedule.

The work of the Human Genome Project has allowed researchers to begin to understand the blueprint for building a person. As researchers learn more about the functions of genes and proteins, this knowledge will have a major impact in the fields of medicine, biotechnology, and the life sciences.

HINT

Human Genome Project has broken the genetic code for what each chromosome controls and is responsible for. For example, chromosome pair 11 is in charge of blood typing, the shape of your fingernails, and whether or not you have freckles.

CLONING

How does it work? The basic idea of **cloning** is simple: copying biological stuff. Human identical twins, plants grown from a clipping, and freshwater sponges are all naturally occurring biological copies—clones.

There are different ways to clone:

- **Reproductive cloning**—for making an animal or plant with the same DNA in the nucleus as another animal or plant
- **Therapeutic cloning**—for making a reserve of "spare parts" of cells with the same DNA in the nucleus as a particular human or animal
- **Recombinant DNA cloning**—for making many copies of a gene often used in scientific research

DNA fingerprint

Since your DNA is found in all of your body cells, scientists can take a sample of your hair, blood, or even skin and map out your genetic sequence. This is the most reliable form of identifying a person from evidence left at a crime scene. Let's learn about how they do it.

The chemical structure of everyone's DNA is the same. The only difference between people (or any animal) is the **order of the base pairs**. There are so many millions of base pairs in each person's DNA that every person has a different sequence.

Using these **sequences**, every person could be identified solely by the sequence of their base pairs. However, because there are so many millions of base pairs, the task would be very time-consuming. Instead, scientists are able to use a shorter method because of repeating patterns in DNA.

These patterns do not, however, give an individual "fingerprint," but they are able to determine whether two DNA samples are from the same person, related people, or nonrelated people. Scientists use a small number of sequences of DNA that are known to vary a great deal among individuals. They then analyze those results to get a certain probability of a match. The test produces a gel display showing the genetic expressions for all samples tested. Notice how lines are formed from each sample tested. The sample that has the same placement of markers is the match. So in this sample, suspect B matches the sample recovered from a crime scene.

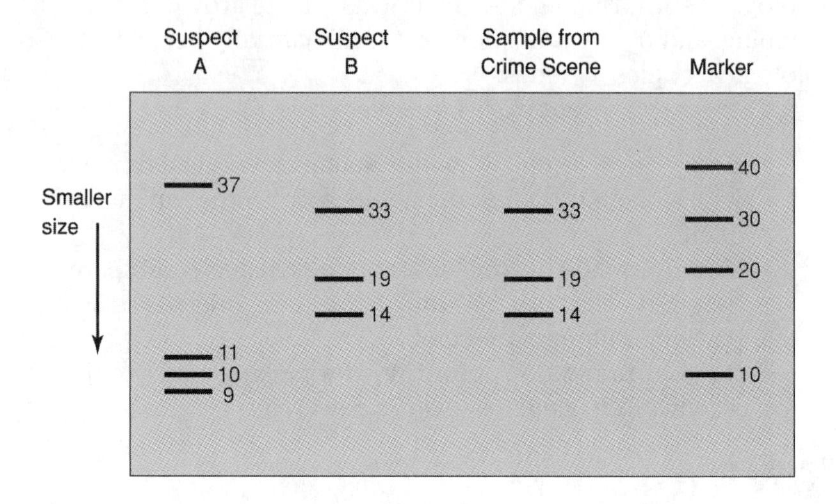

To decide who is related, look for the most similarities.

These genetic tests are used to prove:

- Paternity and maternity
- Criminal identification and forensics
- Personal identification

Acquired characteristics

Some characteristics come from our interactions with the environment and cannot be inherited by the next generation. If you fall and get a bad cut that leaves a scar, you would not pass the scar onto your offspring.

Do you know how to jump rope, ride a bike, tap dance, read, play an instrument, or sing. These are not traits from your parents; these are called **learned** or **acquired behaviors**. You cannot pass these on to your offspring by your genes. You could teach them how to do each, but they are not born knowing how to do them.

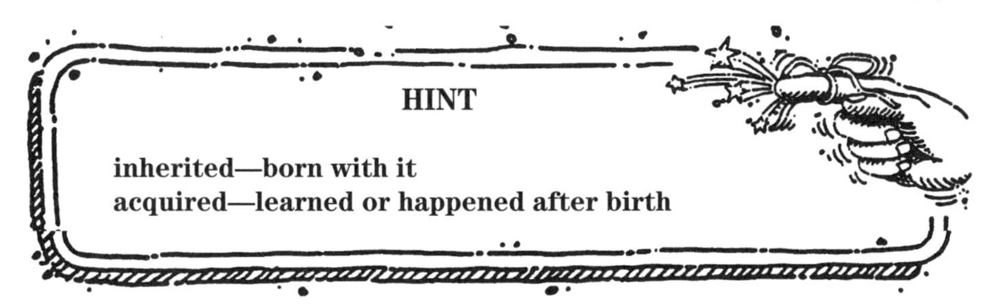

HINT

inherited—born with it
acquired—learned or happened after birth

BRAIN TICKLERS
Set # 32

1. Blood typing is determined by (*circle one:* genotype or phenotype).
2. Creating a copy of an organism is known as __________.
   a. cloning
   b. genome
   c. fingerprint
   d. acquired traits
3. This is the result of the analysis of a DNA sample found at a crime scene. Which subject was at the crime scene? How do you know this?

Suspect A	Suspect B	Sample from Crime Scene	Marker
37	33	33	40
11	19	19	30
10	14	14	20
9			10

Smaller size ↓

4. Explain why the Human Genome Project is thought to be so important to the future of medicine.
5. A woman was born with a large nose; she had plastic surgery to make her nose smaller. Will her children inherit her large nose or her new smaller nose? Explain your answer.

(Answers are on page 179.)

WRAPPING UP

- Genetics is the branch of science that studies how hereditary information is passed on from one generation to another in a family.
- Genes carry the hereditary information for an organism. The information is called a trait. The genes are made up of DNA and are located in the nucleus of all cells.
- DNA is packaged into structures called chromosomes.
- There are forty-six chromosomes in the human. Each chromosome is responsible for different traits such as height, hair color, skin color, shape of eyes, nose, and so on.
- Gregor Mendel is the father of genetics; he made his discoveries studying characteristics of pea plants.
- DNA replicates by splitting in half in a zipper fashion.
- The picture of the human chromosomes lined up in pairs is called a karyotype.
- Genetic expression is known as a dominant or recessive trait.
- The genotype is the genetic makeup; the phenotype is the physical appearance of the trait.
- Genetics is a matter of probability. This probability is determined by performing a Punnet square of the parents' genotype for a trait.
- The Human Genome Project mapped out a blueprint for the function of each gene and protein in the human body.
- DNA fingerprinting describes how every person's DNA is unique. DNA is being used in court cases and paternity cases.

BRAIN TICKLERS—THE ANSWERS

Set # 30, page 165

1. Gregor Mendel
2. Fruit flies and pea plants have very short life cycles so researchers can see many generations and observe how traits are passed down.
3. a. alleles
4. genetics
5. nucleus

Set # 31, page 172

1. A, C, T, G (in any order)
2. Genotype is the genetic makeup, whether the alleles are dominant or recessive (e.g., TT—homozygous dominant). Phenotype is the physical appearance, what the person or organism looks like (e.g., this person would be tall).
3. This is a girl with Trisomy 21 (Down syndrome).
4. 100 percent of offspring would be tall—heterozygous/hybrid.

	T	T
t	**Tt**	**Tt**
t	**Tt**	**Tt**

Set # 32, page 177

1. genotype
2. a. cloning
3. Suspect B was at the crime scene. This is proven because their gel markers match the sample found at the crime scene.

4. Scientists are now able to isolate individual chromosomes and study the different traits for each. As a result, new medicine, genetic treatments, and possibly cures for genetic disorders can happen in the future.
5. The children would inherit the big nose, because the surgery didn't change her DNA (genotype); it only changed her phenotype (appearance).

CHAPTER EIGHT

Ecology

Ecology is the science that studies the relationships between organisms and their environments.

What would humans need to survive on the moon? Would you be able to re-create all of the parts of the earth's surface so that living organisms could survive? What should be included? If you made a list of all these things, you would have to think about the interactions both living and nonliving have together. These interconnected relationships are called **ecosystems**. An ecosystem is the natural environment consisting of all plants, animals, and micro-organisms (biotic—living) in an area functioning together with all of the nonliving physical (abiotic—nonliving) factors of the environment.

Here is an example of a pond ecosystem.

When we study the environment, two types of factors are discussed; they are abiotic and biotic.

Abiotic factors are not living but have an effect on living things and help them live in a certain environment. Water, soil, light, temperature, sun, and rain are abiotic factors.

Biotic factors are living organisms in the environment. Birds, plants, insects, and animals are biotic.

A **biosphere** is defined as the region of the earth that contains all living organism.

There was a long-term research project going on in the desert of Arizona called the Biosphere. Scientists created a glass dome that had to contain all of the living organisms needed so that humans could survive in it without taking any resources from the outside. Anything living in the dome could only get its air, water, and food from within the dome. The scientists who lived in this dome found that they needed a very large number of plants in order to have enough oxygen for animals to survive.

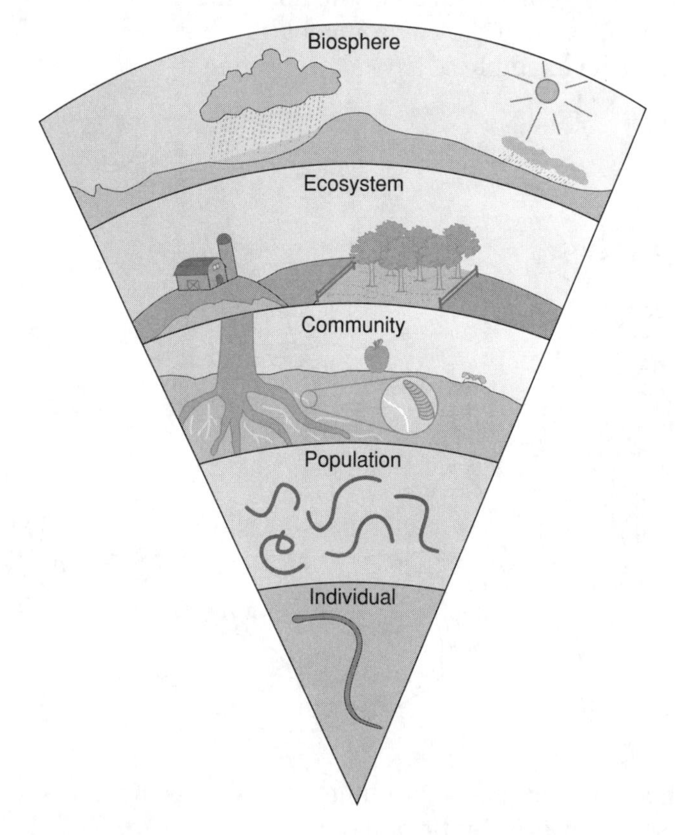

Populations include individual organisms of the same species that are living in the same place and that can produce young.

Community is made up of the population that lives and interacts with populations of other organisms; they depend on each other for food, shelter, and other needs.

Ecosystem is the biotic community and abiotic factors that affect it.

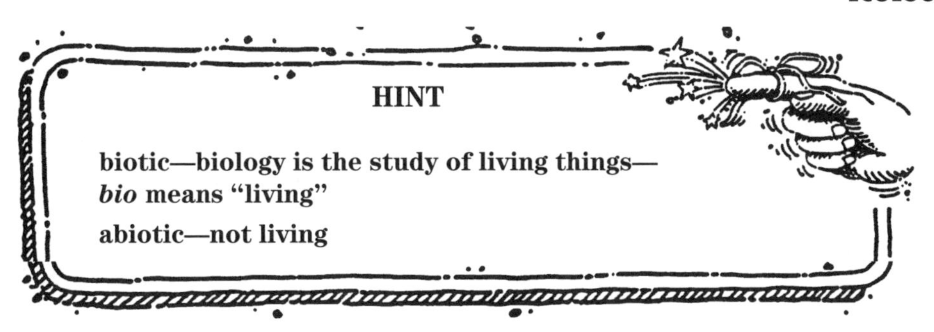

Within our ecosystem, there are **limiting factors** that have an effect on how living things live. Limiting factors are situations that can have an effect on how other organisms survive.

If there are too many organisms living in the same area, they can run out of space to live comfortably. **Population density** has an effect on the survival of organisms. When there isn't enough living space, mates, or resources such as food or shelter and if there is competition with other organisms for any of the basic needs, then the population will begin to die off.

Carrying capacity is the level at which a population is able to survive or reproduce. It is the largest number of individuals an environment can support and maintain for long period of time. Think of your school. There are a maximum number of students that the school can handle safely for the entire year. This would be considered the carrying capacity.

There are many interactions between communities.

Symbiosis is the relationship between two or more different species. To be more specific there are different types of symbiotic relationships:

> **Mutualism** occurs when organisms live together and both species benefit. Bees fertilize the flower and use the nectar to make honey. They both benefit.
>
> **Commensalism** occurs when one organism benefits and the other is not harmed. Remoras attach to the shark, and they feed off the shark's leftover foods. They do not harm the shark.
>
> **Parasitism** occurs when one species benefits while the other is harmed. An example is tapeworms. The host is the organism that the parasite lives off. Tapeworms can live in an animal's intestine. The worm lives off the animal, but it hurts the host animal and in some cases can kill it.

Competition is the struggle between organisms to survive as they attempt to use the same limited resources.

Predation occurs when one organism kills another for food. The **predator** is the organism that kills another for food. The **prey** is the organism that is killed as a form of food. There are many predator/prey relationships: snake and rat, lion and gazelle, and hawk and mouse.

Scavengers are animals that feed off the bodies of dead organisms. You may be familiar with the vulture; it is a large bird that lives mostly in hot regions and feeds off dead animals.

Predators **adapt** so that they can catch more prey. Bats, for example, have the ability to see in the dark. Prey adapts by developing colors that allow them to camouflage against their environment so that they will not be easily seen by the predator.

BRAIN TICKLERS
Set # 32

1. The biosphere includes __________.
   a. small populations
   b. scavengers only
   c. all biotic factors
   d. all biotic and abiotic factors
2. A __________ is a species that benefits from its host.
3. The symbiotic relationship where both organisms benefit is known as __________.
4. Describe how plants and animals, including humans, depend upon each other and the nonliving environment.

(Answers are on page 200.)

FLOW OF ENERGY

In order for anything to move, grow, or change, it needs energy. Energy is very valuable. The **sun** is the ultimate source of energy for all life and physical cycles on earth. The sun's energy is transferred on earth from plants to animals through the **food chain**. Even though plants are able to make their own food by **photosynthesis**, they need sunlight and water to do so.

Plants play the most important part in the cycle of nature. Without plants, there could be no life on earth. They are the primary **producers** that sustain all other life forms. A producer is an organism that can make its own food. Animals are not able to make their own food; they depend directly or indirectly on plants for their supply of food. All animals and the foods they eat can be traced back to plants and to sunlight as its original form of energy. Other organisms eat plants; they are called **consumers**. Energy gets passed onto other organisms for them to use and transfer on through the sequence known as the **food chain**.

Animals in an ecosystem are classified into three different groups based on what they eat. Some animals eat only plants, others eat only other animals, and then some like humans eat both plants and animals. Try and figure out why they got the names they did:

- **Herbivores** are animals that get their energy from eating plants, and only plants. Many herbivores have special digestive systems that let them digest all kinds of plants, including grasses. Herbivores need a lot of energy to stay alive. Many of them, like cows and sheep, eat all day long.
- **Omnivores** are animals that eat both plants and animals. They can be either the primary or secondary consumer.
- **Carnivores** are animals that eat only meat. Their digestive systems are not designed to get nutrients or energy from plants. The carnivore is found at the higher end of the food chain as the secondary consumer or tertiary consumer.

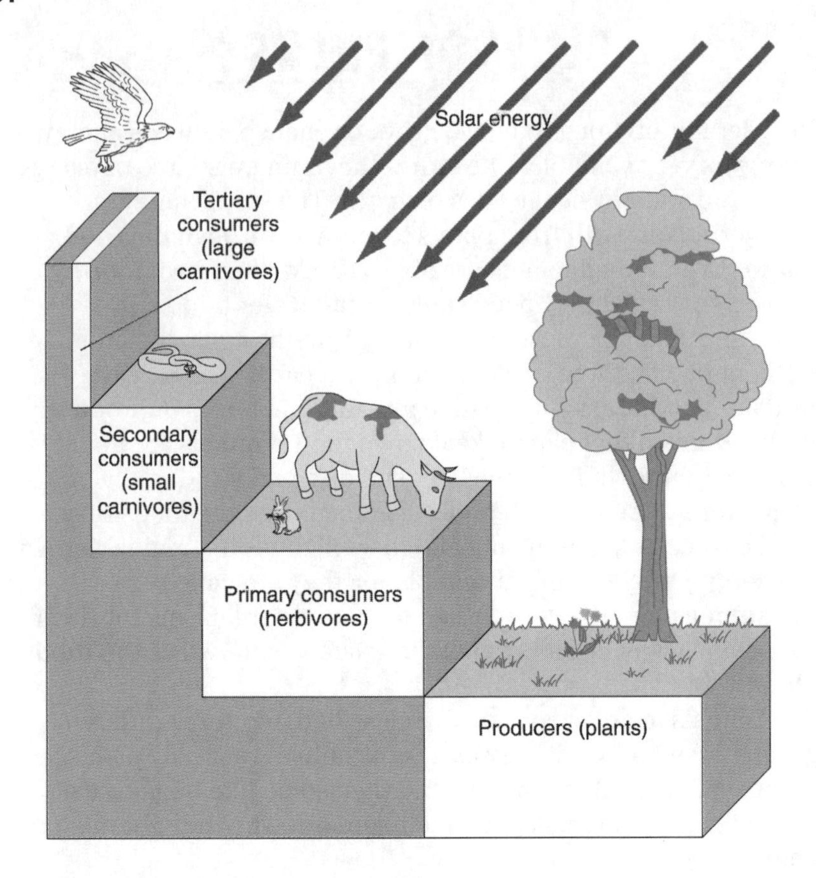

Three terms are used when discussing a food chain:

- **Producers** are living things that change the sun's energy to make their own food. Green plants do this by photosynthesis.
- **Consumers** are living things that feed on others.
  - A primary (first) consumer is a herbivore that eats producers (plants).
  - Secondary (second) consumer eats the primary consumer.
  - Tertiary (third) consumer eats the secondary consumer.
- **Decomposers** are living things that break down dead organisms and recycle their nutrients into the soil. They help in plant growth.

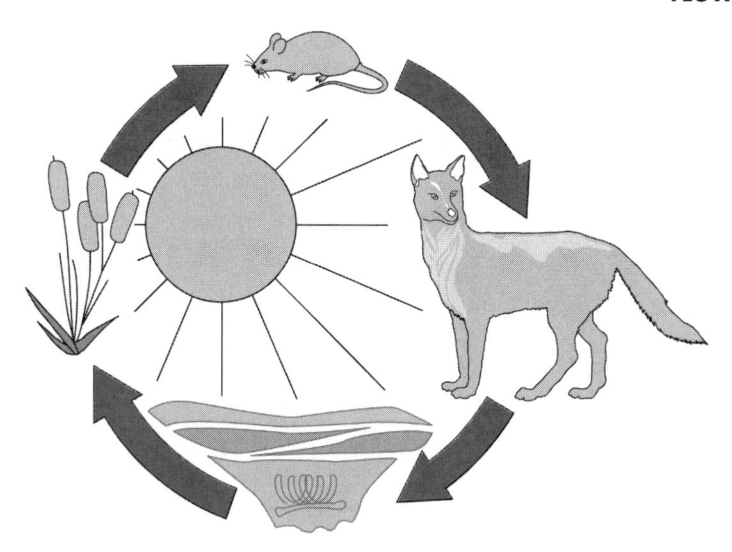

The plant uses sun and nutrients from decomposed waste in the soil to grow, the mouse eats the plant, the fox eats the mouse, the fox's waste is decomposed by bacteria in the soil that creates nutrients for the plant and the whole cycle continues.

Life doesn't always work in a single way. When food chains crisscross it creates a **food web**.

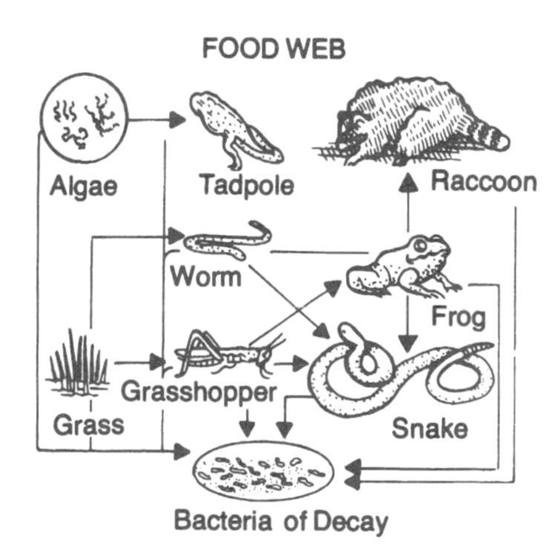

The amount of energy that is transferred from one organism to the next varies in different food chains. Generally, about *10 percent* of the energy from one level of a food chain makes it to the next level. The wider part of the triangle represents the greatest amount of energy.

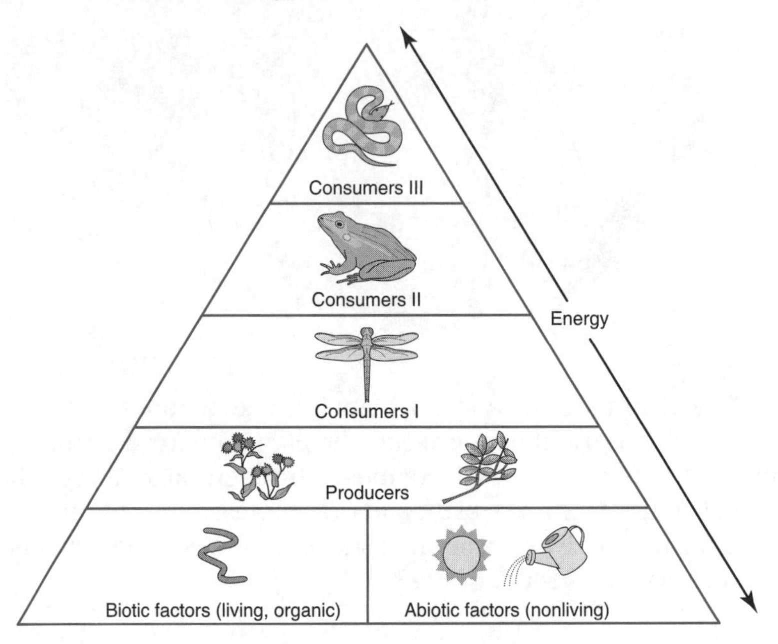

BRAIN TICKLERS
Set # 33

1. The __________ is the ultimate source of energy for all life and physical cycles on earth.

Questions 2–5 relate to the diagram below:

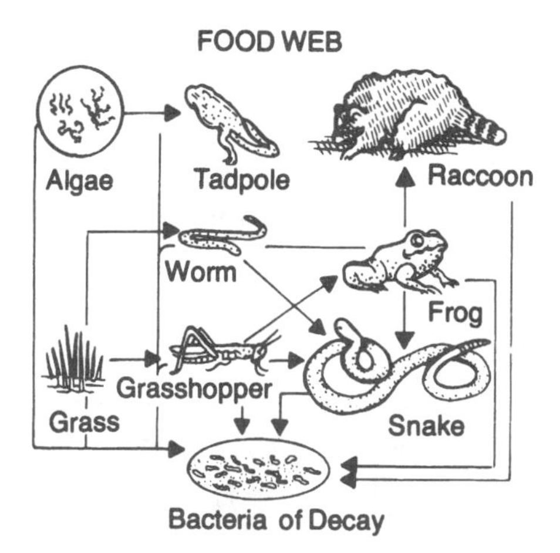

2. This picture represents __________.
   a. a food web
   b. ecosystems
   c. a food chain
   d. decomposers
3. The cactus is a __________.
   a. producer
   b. consumer
   c. decomposer
   d. predator
4. The butterfly is a __________.
   a. producer
   b. consumer
   c. decomposer
   d. predator
5. The snake is a __________.
   a. producer
   b. primary consumer
   c. decomposer
   d. predator

Answer questions 6–8 based on this drawing.

6. Create a food chain using animals that you see in the drawing.
7. Identify a predator.
8. Identify a prey.

(Answers are on page 200.)

CYCLES IN OUR ECOSYSTEM

Water cycle

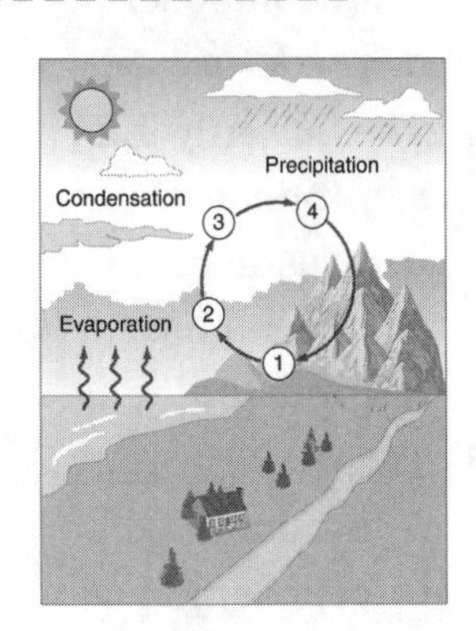

The water that is on the earth today is believed to be the same that was on the earth a million years ago when the dinosaurs roamed the earth. Water is constantly moving through the water cycle: evaporation, condensation, and precipitation. See diagram. 1 = runoff, 2 = evaporation, 3 = condensation, and 4 = precipitation.

Nitrogen cycle

Nitrogen is an important element needed to make proteins. Living organisms would not be able to survive without proteins.

In the atmosphere, there is a large amount of nitrogen gas that needs to be changed into a usable form. Nitrogen has to be combined with other elements in the process called nitrogen fixation. Nitrogen is converted by bacteria for use by plants.

Carbon and oxygen cycle

The process by which carbon and oxygen are recycled is linked. Producers, consumers, and decomposers play roles in recycling carbon and oxygen. Producers release oxygen via photosynthesis. Carbon dioxide is taken in by producers and is used to help the plant make food while organisms eat food and give off carbon dioxide.

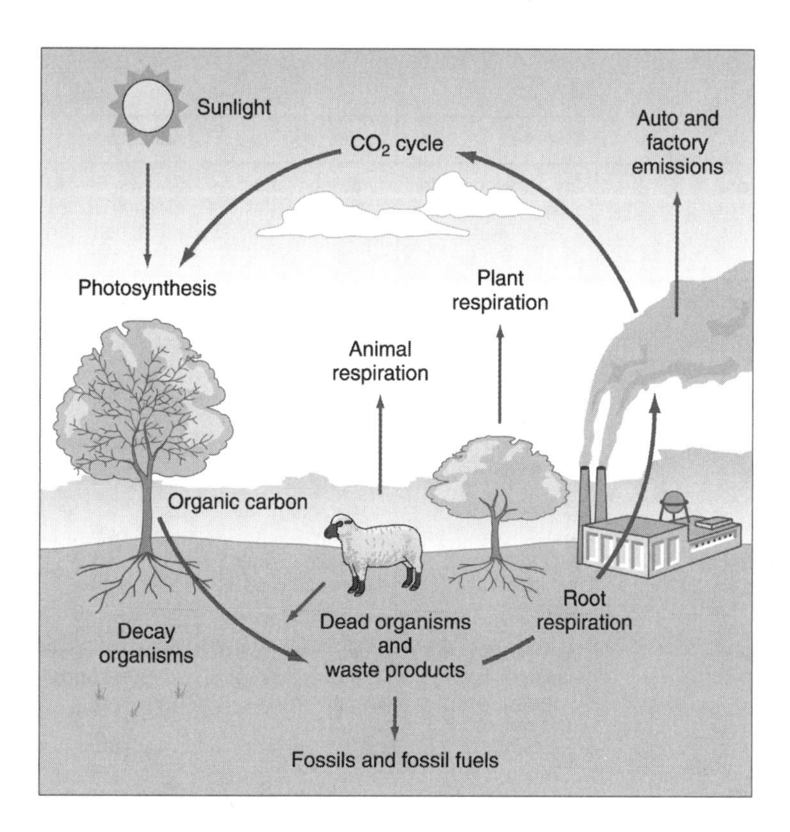

Succession

Primary succession is the first community of organisms to move into a new environment; they are also referred to as a pioneer community.

Secondary succession is a community that comes about after a fire or replaces an area where there once was a building or parking lot.

Climax community is the end product of ecological succession; an example would be a mature full-grown forest.

Pond succession occurs when an area that once had a body of water fills in over time and a full terrestrial area emerges. Think of it, there are areas where fossils of fish are found miles from water. There probably was a body of water in that area at some time in the past and succession occurred.

A **biome** is a geographic area defined by factors such as plant structure—vegetation, climate, and the animals that live in that climate.

Biome	Climate	Vegetation	Animals
Tundra	Dry, very long cold seasons	Growing plants like mosses, heaths, and lichen	Shrews, hares, rodents, wolves, foxes, bears, and deer
Taiga	Moist, cold season, get snow, summers can be in the 70s	Pine trees, evergreens	Fur-bearing mammals, migratory birds
Temperate deciduous forest	Broadleaf trees, creates canopy in summer, trees lose leaves in winter	Humid subtropical, lots of rain, hot, sultry summers, mild to cold winters, eastern side of continents	Many different kinds—birds, reptiles, insects, mammals
Tropical rain forest	Very moist and humid, always warm	Numerous plants, thick vegetation	Numerous birds, snakes, reptiles
Desert	Very dry, long, hot summers, large fluctuation in temperature during day	Plants that survive in dry areas, cactus, sporadic trees	Few mammals, reptiles, burrowing animals
Grasslands	Hot summer, cold winter, dry, transition between desert and humid climate	Low-lying plants and grasses, no large plants due to dry conditions	Grazing animals, herds of large animals, small rodents that spend time in holes underground like gophers

BRAIN TICKLERS
Set # 34

1. Which cycle recycles decaying matter and waste for plants to use?
2. Explain how the water cycle works.
3. Is water made?
4. Explain how succession is an important part of life on earth.
5. Choose two biomes and explain how the organisms and vegetation are suited for their climate.

(Answers are on page 201.)

RESOURCES

Earth's **natural resources**—parts of the environment used by living organisms for food, shelter, and other needs—are naturally occurring substances that are considered valuable in their relatively unmodified (natural) form. A natural resource's value rests in the amount of the material available and the demand for it. Examples of natural resources are: cotton, gasoline, electricity, air, water, and food.

Renewable resources are natural resources that are recycled or replaced by ongoing natural processes. Examples are wind, sun, and geothermal resources.

Nonrenewable resources are natural resources that are available in limited amounts and are not replaced by natural processes in a short period of time. Examples are oil, nuclear resources, coal, and natural gas.

Energy causes things to happen around us. Look out the window. During the day, the sun gives out light and heat energy. At night, street lamps use electrical energy to light our way. When a car drives by, it is being powered by gasoline, a type of stored energy. The food we eat contains energy. We use that energy to work and play. The definition of **energy** is the ability to do work.

Energy can be found in a number of different forms. It can be chemical energy, electrical energy, thermal energy (heat), radiant energy (light), mechanical energy, and nuclear energy. The forms of energy are electricity, geothermal energy (heat energy that comes from the earth), fossil fuels (coal, oil, and natural gas), hydro power and ocean energy, nuclear energy, solar energy, wind energy, and biomass energy (energy from plants).

HINT

Renewable—easily made in a short period of time or available everywhere

Nonrenewable—takes a very long time to form

Recycling is when you take old materials and make them into new products. To prevent the waste of potentially useful materials, reducing the consumption of fresh raw materials, reducing energy usage, reducing air (from incineration) and water (from landfills) pollution reduces the need for the usual waste disposal. Recycling is a key concept of modern waste management.

Recyclable materials or recyclables may come from home, business, or industry. They include glass, paper, metal, textiles, and plastics. These materials are sorted, cleaned, and reprocessed into new products bound for manufacturing. Recycling in the end is done to reduce the amount of energy that is needed to create new things and the buildup in the amount of waste in landfills.

Energy conservation is the practice of decreasing the quantity of energy used. It may be achieved through efficient energy use or reduced consumption of energy services. Individuals and organizations that are direct consumers of energy may want to conserve energy in order to reduce energy costs. Industrial and commercial users may want to increase efficiency and thus maximize profit.

Energy conservation reduces the energy consumption and energy demand per person, and thus reduces the growth in energy supply needed to keep up with population growth. This reduces the rise in energy costs and can reduce the need for new power plants and energy imports. By reducing emissions from cars or factories, energy conservation is an important part of lessening climate change.

Pollution is the introduction of contaminants into an environment. These contaminants cause instability, disorder, harm, or discomfort to the physical systems or living organisms. Pollution can take the form of chemical substances or energy, such as noise, heat, or light energy. Pollutants can be foreign substances or energies, or they can be naturally occurring. Fossil fuels give off into the environment chemicals that result in smog and harmful gases that are considered pollutants.

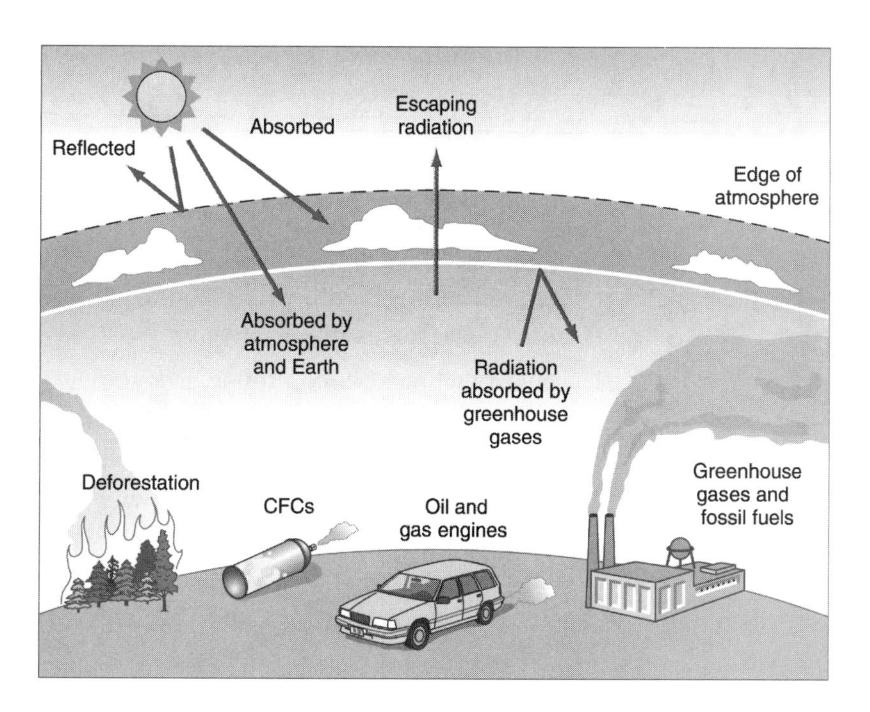

GREENHOUSE EFFECT

There is much information coming out about the **greenhouse effect** and how it is affecting our climate.

The greenhouse effect is real and helps to regulate the temperature of our planet. Keeping steady temperatures in an environment is essential for life on earth. Because an increased amount of gases are released into our environment, a hole is being formed in the protective layer surrounding the earth called the ozone layer. And because more heat is getting in through the ozone layer and certain gases are being absorbed in the atmosphere, causing them to be trapped like a greenhouse, we use the name greenhouse effect. The temperature of the earth has been increasing due to this process.

The earth's (global) surface temperatures have increased since the late-nineteenth century. This increase has gotten worse in recent years. Seven of the eight warmest years on record have occurred since 2001 and the ten warmest years have all occurred since 1995. Due to global warming, the ice caps are melting faster, crops are not growing to their full potential due to increased heat, and ecosystems are being altered by this increased heat.

BRAIN TICKLERS
Set # 35

1. Explain why it is important to find alternative forms of energy.
2. Explain what Reduce, Reuse, Recycle means and how it would help the environment.
3. Name one harmful pollutant and why it is harmful.

4. The greenhouse effect got its name because _________.
   a. pollution has increased the plants on the earth
   b. sunlight has increased pollution on the earth
   c. harmful gases have caused average temperatures on the earth to rise
   d. harmful gases have decreased the temperature on the earth

(Answers are on page 201.)

WRAPPING UP

- Ecology is the science that studies the relationships between organisms and their environments.
- Pollution is the introduction of contaminants into an environment. These contaminants cause instability, disorder, harm, or discomfort to the physical systems or living organisms.
- Recycling and energy conservation is essential for the long-term harmonious survival of our ecosystem.
- The greenhouse effect occurs when heat gets in through the ozone layer and is absorbed by certain gases in the atmosphere, causing them to be trapped as though they were in a greenhouse. The temperature of the earth has been increasing due to this process.
- A geographic area defined by the vegetation, climate, and animals that live in that climate is called a biome. Tundra, taiga, deciduous forest, desert, and arctic are examples of biomes.
- Food chain, food web, water cycle, nitrogen cycle, oxygen cycle, and succession are all examples of how the ecosystem constantly changes and supports living organisms.
- Living organisms in our ecosystem can be sorted into four categories based on the food chain: producers (plants), primary consumers (eat plants), secondary consumers (eat primary consumers), and decomposers (break down dead materials).

BRAIN TICKLERS—THE ANSWERS

Set # 32, page 186

1. d. all biotic and abiotic factors
2. parasite
3. mutualism
4. Answers may vary. Plants depend on the sun for energy and on water and the soil for nutrients to survive and to perform photosynthesis. Animals need plants for food and energy. Plants give off oxygen for animals to breath in. Animals give off carbon dioxide for the plants to take in. Humans need plants and animals for nutrition. Some animals use nonliving materials for shelter.

Set # 33, page 190

1. sun
2. a. a food web
3. a. producer
4. b. consumer
5. d. predator

Answers may vary for questions 6–8.

6. grass → rabbit → bird

   grass → fly → frog → snake

   tree → squirrel → bird

   water plant → fish → snake → bird
7. bird or snake
8. rabbit, frog, or fish

Set # 34, page 195

1. nitrogen cycle
2. Precipitation—rain, snow, hail, or sleet falls; Evaporation—water returns to the air; Condensation—water droplets form from the air
3. No, water is not made; it changes forms through the water cycle.
4. Answers will vary.
5. Answers will vary.

Set # 35, page 198

1. Fossil fuels are harming the environment and are becoming very expensive. Alternative forms would be better for the environment and cheaper.
2. Limiting the amount of waste reduces the energy used to make items and reduces the amount of waste in landfills.
3. Answers will vary. Car exhaust gives off harmful chemicals that have caused the ozone layer to have holes.
4. c. harmful gases have caused average temperatures on the earth to rise

CHAPTER NINE

Changes over Time

THE HISTORY OF LIFE: LOOKING AT THE PATTERNS

The history of life is constantly changing. We often say it is evolving. **Evolution** is the process of change in the inherited traits of a population of organisms from one generation to the next.

Have you ever visited a historic village where you walk through houses that existed in the 1800s? When you go into a home that was used in the 1800s, the doors are very short compared to what our homes have today. The chairs are much smaller and the beds are much shorter than they are today. People who lived in that time were much shorter than they are today. History also shows us that people did not live as long as they do today. If you lived to your 30s or 40s, you would be considered an elder. Today the average life span of humans is much higher, late 70s or early 80s. Why would this be? Think of the factors that led to the increase in humans' average size and life expectancy; we have better health care, eat more food, and are more conscious of eating healthful food and exercising. Another factor that has an effect on your life expectancy is where you live in the world. Certain countries have lower life expectancy due to disease, poverty, and poor health care.

The central ideas of evolution are that species *change over time* and that different species share common ancestors.

Here, you can explore how evolutionary change and evolutionary relationships are represented in "family trees."

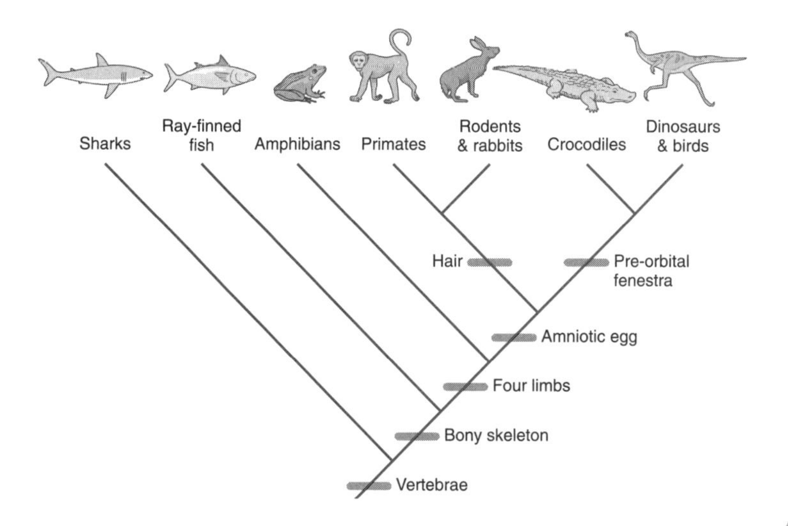

Comparative anatomy supports the theory that animals have a common ancestor. Look at how organisms have "evolved" to help them survive. These are also known as homologous structures.

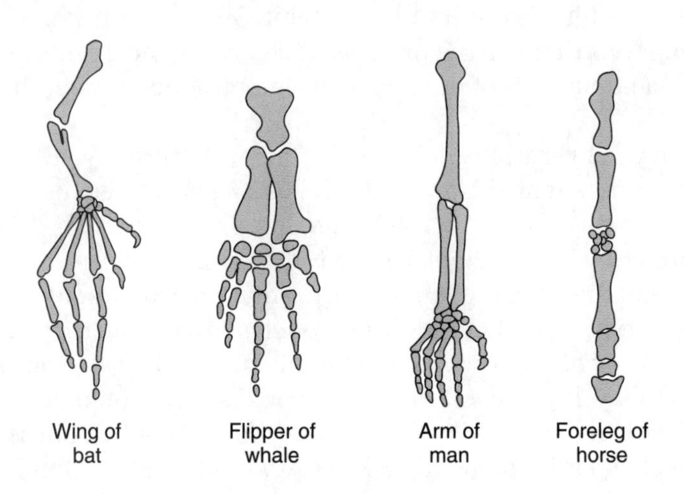

Comparative embryology shows how all animals are very similar as embryos and change as they grow.

Scientists have found that their studies of DNA, embryonic development, and similar body structures support the theory that **evolutionary relationships** among species exist.

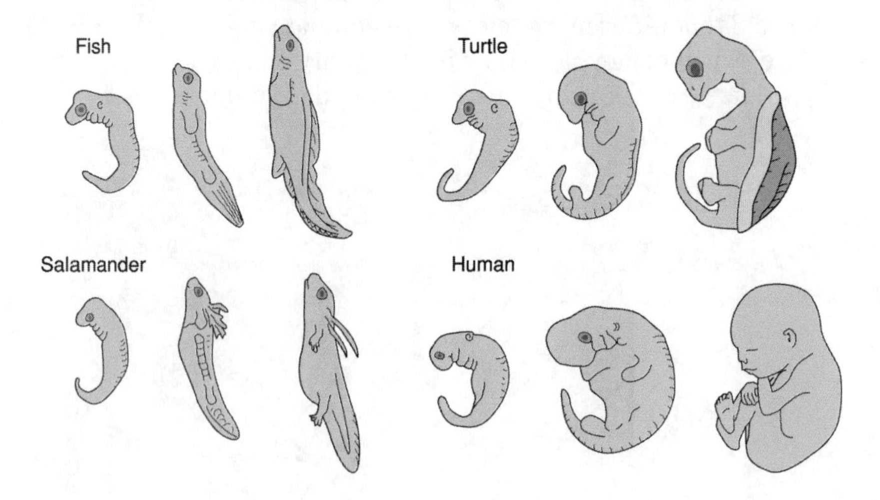

NATURAL SELECTION

Natural selection is the process by which favorable inheritable traits become more common in generations of a population of reproducing organisms, and unfavorable inheritable traits become less common. Natural selection acts on the phenotype, or the observable characteristics of an organism. These phenotypes would be favorable traits that are more likely to help the organism survive and reproduce than those with less favorable traits. Since this favorable trait helps the organisms survive, there are more of them in the gene pool when mating occurs.

Over time, this process can result in **adaptations** that alter organisms to help them thrive in a specific environment and may eventually result in appearance of new species. In other words, natural selection is the mechanism by which evolution may take place in a population of a specific organism.

For example, animals that live in the wild are very good at hiding themselves from predators. Chameleons have the advantage that their skin changes colors to blend into the environment. These lizard-like animals have a greater chance of surviving than those that don't change color. Since the color-changing ones have a better chance of survival, they continue to mate and have offspring born with the same trait.

Natural selection is one of the cornerstones of modern biology. An English naturalist named **Charles Darwin** first gave us the name "natural selection" after his research in the Galapagos Islands. The Galapagos Islands are a string of islands in the Pacific off the coast of Ecuador. The islands have very different environments: one is desert-like, another is like a tropical forest, one is similar to a moonscape with its dark rock and sands, and the others are a combination. Each island has different trees and organisms on it.

Darwin observed the finches on the islands and found that they were structurally different based on which island they lived on. The birds' beaks on the islands that had more fruit trees were designed to get the fruits; other beaks were designed more for getting small insects to feed on. Darwin's theory of evolution did not talk about the genetics but about the *survival of the fittest.*

Different organisms in a population possess different versions of a gene for a certain trait. This genetic variation underlies what it will express as a characteristic or phenotypic trait. A typical example is that certain combinations of genes for eye color in humans give rise to the phenotype of blue eyes. Some traits are governed by only a single gene, but most traits are influenced by the interactions of many genes.

Inheritance of acquired traits

Around the same time as Darwin there was another scientist, **Jean-Baptiste Lamarck**, who gave his theory on how species changed over time. Lamarck's theory was referred to as the inheritance of acquired traits.

For a time, Lamarck's theory was held as an alternative to Darwin's explanation for evolutionary change. Lamarck's famous example is how the giraffe got such a long neck. The classic giraffe analogy offers the best delineation between the two. Lamarck said that the giraffes stretched their necks because the leaves were high on the trees and that resulted in their necks growing. This stretching resulted in the next generation being born with longer necks. Darwin explains that this happened because the giraffes with longer necks were able to get more food and survive to have offspring and thus had offspring that had that long-neck trait.

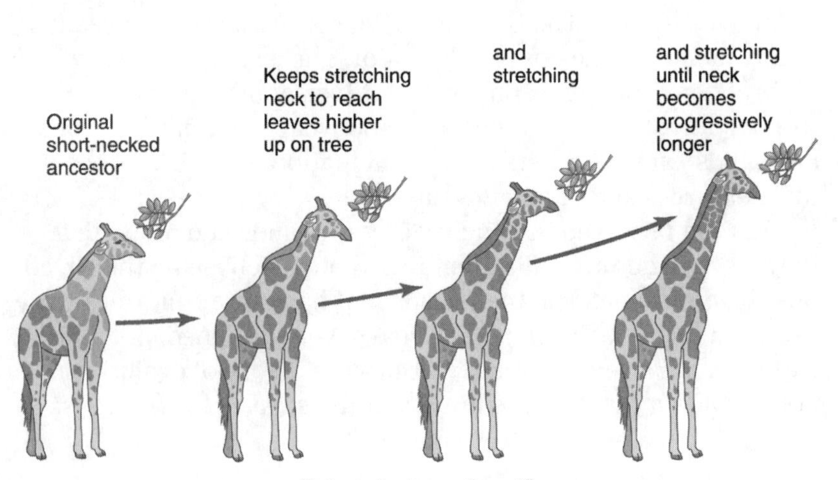

Let's give a modern twist on Lamarck's theory of inherited acquired traits. A woman is born with a less than attractive nose and has surgery to change the shape of her nose. She marries someone who has also had surgery to change the appearance of his nose. When they have a child, is the child born with the new nose? No, the baby would be born with a nose that looks like either one of the parents' noses before the surgery.

Fossils are important for estimating when various species developed. Since fossils are an uncommon occurrence, usually requiring hard body parts and death near a site where sediments are being deposited, the fossil record only provides sparse and intermittent information about the evolution of life.

BRAIN TICKLERS
Set # 36

1. Explain why certain traits are more likely to offer an advantage for survival of an organism.
2. Relate the structure of an organism's skeleton to its ability to survive in a specific environment.
3. A trait that helps an organism survive and reproduce is called a(n) __________.
   a. variation
   b. adaptation
   c. species
   d. selection
4. Give an example, other than that of the giraffe, that would support Lamarck's theory of inherited acquired traits.

(Answers are on page 210.)

WRAPPING UP

- Evolution is the process of change in the inherited traits of a population of organisms from one generation to the next.
- Fossils are important for estimating when various species developed.
- Natural selection is the process by which favorable inheritable traits become more common in generations of a population of reproducing organisms, and unfavorable inheritable traits become less common.
- The central ideas of evolution are that species change over time and different species share common ancestors. This theory is supported by looking at the comparative anatomy and comparative embryology of organisms.

BRAIN TICKLERS—THE ANSWERS

Set # 36, page 209

1. Animals with favorable traits would survive and not be killed by predators, so they could reproduce to carry on positive characteristics.
2. Answers will vary. Hollow bird bones allow birds to fly in air; hair insulates animals from hot or cold; dense root structure allows plants to grow in dense soil; with fins, fish can move in water.
3. b. an adaptation
4. Answers will vary. The monkey's long tail came from the monkey hanging by the tail and it getting longer and longer.

CHAPTER TEN

Health

As you have seen and studied, your body works like a very delicate machine that needs to be taken care of so that it can work and feel its best. In order for humans to grow, have energy, and be healthy they need to eat healthy foods, exercise, and get regular rest.

EAT HEALTHY FOODS

Eating a **balanced diet** is one of the most important steps to having good heath. It is important that you take in the proper foods and nutrients so that your body can grow, have energy, and repair every day.

A balanced diet includes these nutrients:

Grains—bread, rice, pasta, cereal
Vegetables—lettuce, broccoli, carrots, cucumbers
Fruits—apples, bananas, oranges, grapes, pears
Milk—cheese, yogurt, cream, dairy
Meats and beans—hamburger, chicken, eggs, tuna, nuts
Oils and sweets—butter, oil, and candy (all in very limited amounts)

Think before you eat

Not only do you need to make sure that you are eating from the major food groups, you also must be aware of the types of foods you are putting into your body. Fast food and junk food often advertise themselves as having things that are good for you, but they often have chemicals and additives that aren't the best for you. They also can have what is called empty calories. These are useless calories that can cause you to gain weight and result in other health risks.

The **My Pyramid** is an easy way to plan the food that you are going to eat. Each part of the pyramid represents the proper amounts of each type of food that will help you stay healthy. You must also exercise regularly to live a healthy lifestyle.

Be aware of what you eat

All packaged foods sold in the United States must have the nutrition facts printed on the package. This explains how many total calories, calories from fat, nutrients, and vitamins, as well as how much fat and sugar, are found in that food. The number of calories in a serving indicates the amount of energy your body would have to produce to burn off that food. If the total number of calories eaten in a day is greater than those used, fat will develop to store the unused calories, resulting in gaining weight.

Before you eat, read the label.

MACARONI & CHEESE

Nutrition Facts

Start here

Serving Size 1 cup (228g)
Servings per Container 2

Amount Per Serving

Calories 250 Calories from Fat 110

	% Daily Value*
Total Fat 12g	**18%**
Saturated Fat 3g	**15%**
Cholesterol 30mg	**10%**
Sodium 470mg	**20%**
Total Carbohydrate 31g	**10%**
Dietary Fiber 0g	**0%**
Sugars 5g	
Protein 5g	
Vitamin A	**4%**
Vitamin C	**2%**
Calcium	**20%**
Iron	**4%**

Limit these nutrients

Get enough of these nutrients

Quick Guide to % DV

5% or less is low

20% or more is high

* Percent Daily Values are based on a 2,000 calorie diet Your Daily Values may be higher or lower depending on your calorie needs:

Footnote

	Calorie	2,000	2,500
Total fat	Less than	65g	80g
Sat fat	Less than	20g	25g
Cholesterol	Less than	300 mg	300mg
Sodium	Less than	2,400 mg	2,400 mg
Total Carbohydrate		300g	375g
Dietary Fiber		25g	30g

EXERCISE

Make that cardiovascular system work for you! When we run, jump, walk, play sports, or ride bikes, we are **exercising**. Not only is it fun, but it is healthy for our bodies. When we exercise, our body is working hard and moving the nutrients, water, and air throughout our body.

Maintain a healthy weight. Your weight has a direct relationship to your health. People who are overweight give their organs additional work to do. The body has to work harder to move, to pump blood, and to nourish the cells. Extreme cases of being very overweight or obese can have a negative effect on the hair and skin.

GET REGULAR REST

Getting enough rest is also an important part of living a healthy life. Our bodies need time to rest and recuperate from our busy day. While we sleep our bodies have a chance for cells to grow and repair. When we get a good night's sleep, we are able to concentrate better in school and find it easier to complete our work.

PRACTICE GOOD HEALTH HABITS

Being healthy also includes good health habits.

Personal cleanliness

Hand washing and personal cleanliness are necessary to prevent disease and infection.

Many bacteria and germs can be transmitted from touching things and then placing your hands by your face. This is why you need to wash your hands often to remove any possible harmful germs. Think of any public area—picnic tables, benches in the mall, public restrooms—and how many people use them in one day. We all carry germs—some good, some bad. To reduce getting ill or making others ill, we must be socially responsible and keep germs to a minimum. Your body needs to be cleaned regularly to prevent germs and bacteria from growing.

Covering your mouth when you cough or sneeze is another way to prevent passing your germs to someone else.

Oral hygiene

Oral hygiene is also very important. Brushing your teeth in the morning and before going to bed is another important part of having good health. Ideally, you should brush your teeth after every meal. Brushing your teeth removes the food and helps keep teeth strong. Hard candy, juices, and gum can cause a buildup of sugar in between teeth that brings on the cavities. Using dental floss to clean between the teeth will also reduce and hopefully prevent the occurrence of tooth decay. Cavities that are untreated can result in tooth decay and other diseases. Regular visits with your dentist will help you maintain healthy teeth. Remember, once your baby teeth fall out, the adult teeth you have is what you will have for the rest of your life. Take care of them!

Wound care

If you get a cut, it too must be cleaned properly to prevent any bacteria from creating an infection. A cut should be cleaned, treated with an antibiotic ointment, and then covered with a bandage to reduce the chance of its getting infected.

Viral infection

There are certain times of the year when you hear of a virus spreading around school and people getting sick. A **virus** is a submicroscopic infectious agent that is unable to grow or reproduce outside a host cell. A **host cell** is a cell that will act as the home and food for the virus. Each viral particle, or **virion**, is made up of genetic material (DNA or RNA) within a protective protein coat called a capsid. The capsid shape varies depending on the type of virus. Viruses infect cellular life forms.

Biologists debate whether or not viruses are living organisms. Some consider them nonliving because they do not meet the criteria of the definition of life; viruses do not have cells and cannot survive on their own.

Viruses have genes and evolve by natural selection. Viral infections in human and animal hosts usually result in an immune response and disease. Antibiotics have no effect on viruses, but antiviral drugs have been developed to treat life-threatening infections. **Vaccines** that produce lifelong immunity can prevent viral infections.

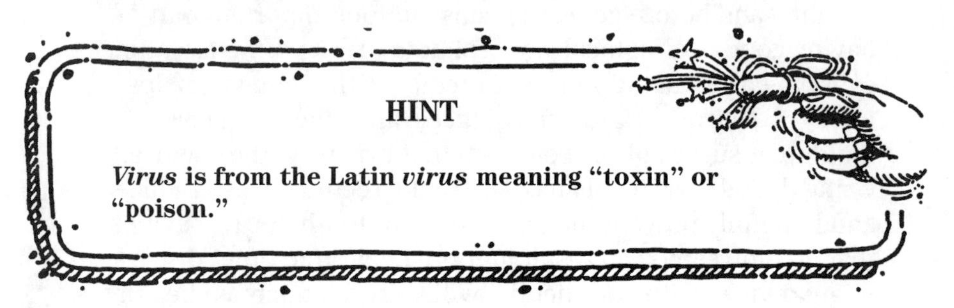

HINT

***Virus* is from the Latin *virus* meaning "toxin" or "poison."**

AVOID HARMFUL SUBSTANCES

You have only one chance to live a healthy life style. The best way to do this—in addition to eating healthy, exercising, and getting rest—is to avoid (do not take or try) harmful substances.

Smoking tobacco or illegal drugs is very harmful to your body. When a person smokes, the tobacco releases chemicals that enter your lungs. Because the lungs have to work harder to remove the chemicals, your body has to work harder to get the air needed to survive. The chemicals from smoking can also cause very serious diseases such as lung cancer. Your skin and teeth turn an unhealthy color as well, due to smoking. Look at the difference between a

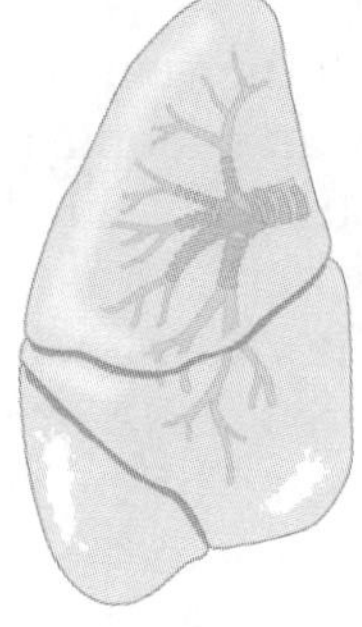

Non-smoker

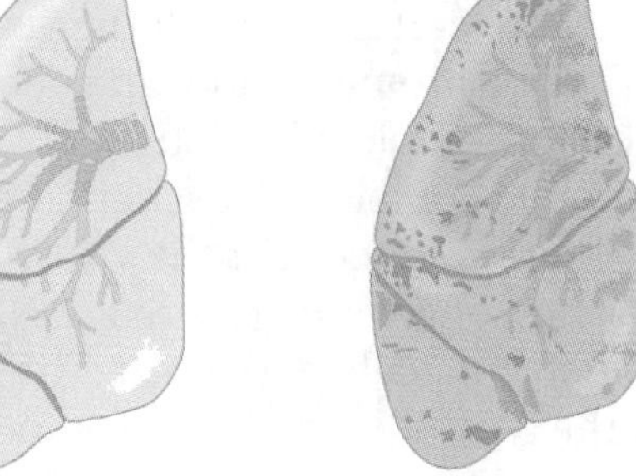

Smoker

healthy lung and a smoker's lung. The lung is responsible for filtering out oxygen for use for all life processes. Think of how much damage can occur if the body does not get enough oxygen.

Drinking **alcohol** and/or taking illicit (not prescribed) **drugs** are also very harmful to your body. Using these substances can distort your mind, which can lead to destructive decisions. People who are under the influence of drugs or alcohol have impaired reaction time and may have blurred vision, loss of memory, and other disorders due to these substances. To dispose of the alcohol and drugs, the blood takes these substances to the liver, which is the organ that cleans the blood. The liver has to work harder to remove these foreign substances. It can result in your body developing some serious illnesses. These substances also can lead to people developing addictions. **Addiction** is a state in which the body relies on a substance for normal functioning. When this substance is removed, it can cause **withdrawal**. A number of long-term physical problems and diseases can result from addiction.

BRAIN TICKLERS

Set # 37

1. Name two ways that food is used by the body.
2. To avoid getting germs and illnesses from other people, you should __________.
   a. eat bread every day
   b. exercise weekly
   c. wash your hands frequently
   d. go to bed on time
3. Puberty occurs during __________.
4. Explain what oral hygiene is and why it is important.

5. The food pyramid shows us __________.
   a. how not to eat
   b. what foods make a balanced diet
   c. we don't need to exercise
   d. what foods to avoid
6. List the six nutrients that are needed by the body.
7. An example of a poor health habit is __________.
   a. eating candy for dinner
   b. washing your hands after using the bathroom
   c. playing at the park every afternoon
   d. getting eight hours of sleep at night
8. Describe how exercise is beneficial to having a healthy body.
9. Explain how cigarettes can affect your body.
10. Explain how a virus survives. How can you prevent some viral infections?

(Answers are on page 221.)

WRAPPING UP

- In order for humans to grow, have energy, and be healthy, they need to eat healthy foods, exercise, and get regular rest.
- Eating a balanced diet is one of the most important steps to having good health. It is important that you take in the proper foods and nutrients so that your body can grow, have energy, and repair everyday. The six nutrients needed by the body are grains, vegetables, fruits, milk, oils, and meats and beans.
- Drugs and alcohol have a harmful effect on the life span of a human.

BRAIN TICKLERS—THE ANSWERS

Set # 37, page 219

1. Food is used for energy production, nourishment, growth, and repair of cells.
2. c. wash your hands frequently
3. adolescence
4. Good oral hygiene is brushing your teeth and flossing. Sugar and bacteria can build up between teeth and cause the tooth to develop cavities or decay.
5. b. what foods make a balanced diet
6. The body needs grains, vegetables, fruits, oils, milk, and meat and beans.
7. a. eating candy for dinner
8. Exercise provides the body with good blood flow for gas exchange. It also allows the body to burn calories and prevent obesity.
9. Cigarette smoke hinders the exchange of gases in the alveoli of the lungs. It gives off harmful chemicals that have been known to cause diseases and birth defects.
10. A virus has to enter a host cell to survive. Once there, it can replicate its DNA and create additional virons. Vaccinations are given to help people fight a virus if it is introduced.

INDEX

D

E

R

S